Literatures, Cultures, and the Environment

Series Editor

Ursula K. Heise
University of California
Dept of English
Los Angeles, California, USA

Literatures, Cultures, and the Environment focuses on new research in the Environmental Humanities, particularly work with a rhetorical or literary dimension. Books in this series explore how ideas of nature and environmental concerns are expressed in different cultural contexts and at different historical moments. They investigate how cultural assumptions and practices, as well as social structures and institutions, shape conceptions of nature, the natural, species boundaries, uses of plants, animals and natural resources, the human body in its environmental dimensions, environmental health and illness, and relations between nature and technology. In turn, the series makes visible how concepts of nature and forms of environmentalist thought and representation arise from the confluence of a community's ecological and social conditions with its cultural assumptions, perceptions, and institutions.

More information about this series at
http://www.springer.com/series/14818

Antonia Mehnert

Climate Change Fictions

Representations of Global Warming in American Literature

palgrave
macmillan

Antonia Mehnert
Munich, Germany

Literatures, Cultures, and the Environment
ISBN 978-3-319-40336-6 ISBN 978-3-319-40337-3 (eBook)
DOI 10.1007/978-3-319-40337-3

Library of Congress Control Number: 2016955971

Cover illustration: © Colorbox.de
Cover design: Will Speed

Printed on acid-free paper

This Palgrave Macmillan imprint is published by Springer Nature
The registered company is Springer International Publishing AG Switzerland

Acknowledgements

A lot of work, research and dedication went into this book and its completion wouldn't have been possible without the support of many individuals and institutions. *Climate Change Fictions* is based on my doctoral dissertation submitted to the Ludwig-Maximilians-Universität München in 2014. The Rachel Carson Center and its interdisciplinary PhD program offered the intellectual environment and academic community that influenced great parts of my work. I would like to thank my supervisor, Christof Mauch, for always taking the time to provide advice and critical insights beyond disciplinary confines and for his continuous encouragement and enthusiasm for my project and work. Sylvia Mayer, my 'second' supervisor, helped with valuable feedback on my analyses and guidance to situate my work in the context of ecocriticism. I am particularly thankful to Alexa Weik von Mossner for our exchange of ideas in various stages of my project. Her comments led me to rethink parts of my original argument and brought a new focus to my work. An equally important influence for my first drafts was Mike Ziser. I am extremely grateful to Louis Warren and Mike for the opportunity to spend a couple of months researching at the University of California Davis, to attend its Environments and Societies Workshop Series, and to have the chance to discuss my work with other ecocritics and graduate students in the environmental humanities in the lively Davis University context.

I was lucky to have a great group of young environmental humanities scholars around me at the Rachel Carson Center as well. To engage with other PhD students from very different disciplinary and national backgrounds was an enriching experience and inspired me to also look at my own research topic from various angles. I particularly would like to thank Rebecca Hofmann, Dania Achermann, and Amir Zelinger for their reliable support and trusted collegiality as well as many of other members of the PhD program for not only “hanging in there” but also having fun together. I would also like to thank the program coordinators, Elisabeth Zellmer and Rob Emmett, for their patience and continuous support for our group of doctoral students. Danish visiting scholar, Gregers Andersen, shared my enthusiasm for the topic of climate change fiction and I am thankful to him for sharing theoretical texts as well as his thoughts on books that we both analyzed. I really enjoyed bringing our research together in a co-authored article on German climate change fiction. Two of my dearest colleagues with similar disciplinary backgrounds deserve special mention here. Hanna Straß, thank you for not only being a great conference buddy and co-founder of the postgraduate forum for Environment, Literature and Culture (ELC), but also reader of many of my chapters, friend and hiking partner. Stephanie Siewert, expert on theory and reader with a critical mind, always providing ideas that brought new aspects of my work to light, I couldn't have done this without your encouragement and friendship. I am also indebted to my dear friend, Fran Lawther, who generously gave up her spare time to proofread several of my chapters, and John Grennan who eased out any strange “Germanisms” that may have still been in the text. Needless to say that all remaining mistakes are my own.

A work like this would have been much more difficult to complete without the support of family and friends. I thank them for their faith in me and interest in my work, their understanding for my absence on numerous occasions, their open arms and helping hands when needed. Finally, Damien, I could not be more thankful for everything you've have done to support me along the way. Thank you.

I am grateful to the sdw (Stiftung der Deutschen Wirtschaft) as well as the LMU Graduate Center for their scholarships, which allowed for the necessary financial independence to conduct my research and complete the dissertation. I am also thankful for the conference travel grants provided by the PhD Program of the Carson Center and their support of graduate student activities, such as the extremely valuable and fun writing group with Chris Pastore and two workshops of the postgraduate Forum “Environment, Literature, Culture.” These academic exchanges were not only fruitful for my work, because they allowed me to share various chapters of my work with others, but also because they helped to keep up the motivation during the more difficult times every PhD student goes through. Finally, I thank Prof. Kunow and Prof. Priewe for their initial encouragement to pursue a PhD with an environmental topic.

A part of Chap. 3 appeared in *Cultural Dynamics of Climate Change* edited by Bernd Sommer (Brill 2015) under the title "Back to the Future: Imagining Climate Change Futures in US American Literature." My analysis of Amsterdam's *Things We Didn't See Coming* was originally published as "Riskscapes in Climate Change Fiction" in *The Anticipation of Catastrophe*, edited by Sylvia Mayer and Alexa Weik von Mossner (Winter 2014). Both are reprinted here with full permission by the original publishers. I thank Ursula Heise for her encouragement to publish this manuscript in this series and her advice on the manuscript as well as the anonymous reviewer for her/his helpful remarks and Ryan Jenkins and Paloma Yannakakis for their editorial efforts.

Contents

Introduction: Imagining Climate Change Futures

On November 12th, 2008, *The New York Times* front page read: "Congress has voted to place ExxonMobil … and other major oil companies under public stewardship, with the bulk of the companies' profits put in a public trust … and used for alternative energy research and development in order to solve the global climate crisis."[1] If readers that day let their gaze wander, they were likely to find other headlines such as "Iraq War Ends," "Court Indicts Bush on High Treason Charge" or "National Health Insurance Act Passes." In disbelief of these powerful news, and in search for a verification, they would soon stumble upon the changed slogan in the upper left hand corner of the so familiar looking *The New York Times*, which said "All the News we Hope to Print" instead of the actual "All the News That's Fit to Print." About 80.000 copies of a fake *The New York Times* edition were handed out that day for free to commuters and city-dwellers alike, leaving most of them in a state of happy confusion. Several hours later the authorship of the mock paper was revealed when the culture-jamming activist duo, the Yes Men, admitted their partial involvement in this spoof.[2] Together with a team of professional graphic designers and journalists they had created this issue to provoke critical reflection on the current political situation and provide readers with ideas about what different futures could lie ahead. Some of the readers' generally positive reactions were filmed and later included in the documentary *The Yes Men Fix the World*.

A. Mehnert, *Climate Change Fictions*,
DOI 10.1007/978-3-319-40337-3_1

The Yes Men parodies give voice to the ways cultural practices and their imaginaries defamiliarize patterns of perception, challenge societal expectations and provide room for reflection. While readers typically expect the "news" to provide them with new and accurate information about the state of the world, the Yes Men questioned this implicit reader-media contract by surprising readers with information that was too unlikely to be true. By playing with readers' expectations and confronting them with different imaginaries of what "news" could be, the Yes Men raised awareness to certain problems in society and offered a socio-political critique of them. Whereas the 2008 *The New York Times* issue tantalized readers with positive visions of the future, one year later—just a day before the 2009 United Nations Climate Change Conference in Copenhagen—the Yes Men distributed a special issue of the *New York Post* that peddled a hyperbolic vision of doom announcing the city would soon succumb to quickly changing climatic conditions under the headline: "We're Screwed!"[3] The news stories showed how rising global temperatures would impact New Yorkers should Congress continue to ignore the dangers of global warming. Flooded subway lines topped a list of inconveniences that New Yorkers would face. Altogether, the news snippets provided an idea about the possible effects of climate change on the city and its inhabitants. By crafting either utopian or dystopian narratives in two different media outlets, the Yes Men allowed readers to envision multiple futures, thereby generating a debate about the consequences of contemporary decision-making. As Kathryn Yusoff and Jennifer Gabrys pointedly argue, "Interventions undertaken through aesthetic practices are like an alternating current, drawing on the everyday and common worlds we inhabit, but also revealing what is strange about those worlds" (17). Aesthetic and cultural practices like the Yes Men's can thereby provide the necessary distance to critically reflect on the present socio-political situation and its impact on the environment.

The cultural realm offers a rethinking and reimagining of contemporary environmental problems such as climate change that not only intervenes in current debates but also fundamentally shapes them. Performances like the Yes Men's, but also literature and film, make an important contribution to our understanding of climate change by depicting how future generations might adapt or might fail to adapt to climatic changes. Especially in the case of climate change, with its characteristically prospective dimension, discursive realms that allow readers and viewers to affectively relate to the future are of utmost importance. Moreover, cultural texts pose ethical questions concerning not only the extent of the human impact on the planet but also decisions about who or what is saved or left to die in a climatically changed future. They therefore engage in a "poetics of responsibility," that is, in a discussion about the responsibility humankind has toward its own actions.[4] Also anthropologist Arjun Appadurai has prominently emphasized culture as a social as well as political practice. According to him,

> The image, the imagined, the imagery—these are all terms which direct us to something critical and new in global cultural processes: the imagination as a social practice. No longer mere fantasy (opium for the masses whose real work is elsewhere), no longer simple escape (from a world defined principally by more concrete purposes and structures), no longer elite pastime (thus not relevant to the lives of ordinary people) and no longer mere contemplation (irrelevant for new forms of desire and subjectivity), the imagination has become an organized field of social practices, a form of work … and a form of negotiation between sites of agency ("individuals") and globally defined fields of possibility. (269)

Culture jammers, authors or filmmakers play a vital role in communicating, addressing and possibly redressing the impact of climate change. Further highlighting the importance of literature and texts beyond the aesthetical realm, Sylvia Mayer has pointed out, "Texts … direct our perception, they suggest categories of interpretation and evaluation, they function as premises for subject and identity formation, for the creation of ethical systems" ("Literary Studies" 112).[5] They thus function as more than mere representation. Cultural texts do not only narrativize certain facts or circumstances but also mediate and shape our very reality.

Yet, climate change, this unprecedented, invisible and elusive phenomenon, has so far been predominantly understood through emission scenarios, diagrams, and other lifeless modes of description. This preference for scientific interpretations of the "weather" has not only enforced a division between the scientific and the human realm but also for a long time suppressed other perceptions of global warming.[6] Bryan Wynne correctly points out that the current climate debate prompts us to "ask ourselves whether the intensely scientific framing of the issue, combined as this is with an intensely economistic imagination and framing of the appropriate responses, may engender profound alienation of ordinary human subjects around the globe from 'owning the issue' and thus from taking responsibility for it" (291). In order to understand this unprecedented phenomenon not only in its scientific, but also cultural complexity, it is important to consider it within a broader context of discourses and narratives, which implies an awareness of social and cultural spheres through which climatic changes are brought to the fore.

In this book I turn to a set of contemporary American novels and read them as an intervention in the predominantly scientific and techno-managerial engagement with anthropogenic climate change.[7] Climate change fiction—literature dealing explicitly with anthropogenic climate change—gives insight into the ethical and social ramifications of this unparalleled environmental crisis, reflects on current political conditions that impede action on climate change, explores how risk materializes and affects society, and finally plays an active part in shaping our conception of climate change. It thus serves as a cultural-political attempt and innovative alternative of communicating climate change.

However, cultural narratives of climate change are often disregarded in the public discussion.[8] Commonly, as Timothy Luke elaborates, "'the facts of life' pass into fields of control for disciplines of ecoknowledge and spheres of intervention for their management" (91). In other words, climate change is considered an issue of concern solely for natural sciences and engineering practices. It is thus turned into a problem that is assumed to be solved best with a scientific-technological fix. This kind of thinking originates in the historically-entrenched division of nature and culture in Western societies, which has so far determined who possesses the legitimate power to address issues related to nature and which was epitomized in C.P. Snow's "two cultures" claim.[9] The disciplinary divide based on the firm belief that nature and culture are separate entities has, consequently, ensured a long-lasting dominance of science in climate change discourses.[10]

Yet in the case of climate change, there is a prognostic and therefore imaginative dimension inherent in scientific texts, and especially in emission scenarios, which for this reason, rather polemically, could be called "science fictions."[11] Even though facticity is commonly ascribed to scientific climate change scenarios, neither scientists nor anyone else knows exactly "how such a planetary transformation might affect particular places and individuals, therefore, [it] amounts to a paradigmatic exercise in 'secondhand nonexperience,' envisioning a kind of change that has not occurred before" (Heise, *Sense of Planet* 206). Also climate scientists have pointed to the "narrative" element in science:

> Climate models are our crystal balls to get a glimpse of the future, an age-old dream of humanity. But what they offer is not a *prediction*—rather, it comes as a bundle of scenarios … Natural scientists can in principle not

> predict the future of climate, because it depends on human actions ... That is a matter of choice, not a pre-determined future that can be calculated today. What we can do, however, is calculate scenarios in "what if" style. (Archer and Rahmstorf 126)

The Intergovernmental Panel on Climate Change (IPCC) developed a range of emission scenarios in order to provide plausible narratives on how the future might unfold based on demographic, social, economic and technological developments.[12] Literary critic, Eva Horn, further clarifies that scenarios are not predictions but anticipations of possible future developments.[13] They depict one version of a story, which could be completely different if only some of its parameters are slightly changed (5) and explore possible future realities in the attempt to serve as a means of reflection about current human behavior. Even though the IPCC report on greenhouse gas emission scenarios clearly states that no probabilities are ascribed to the scenarios nor does the report give preference to any single emission path (*Emission Scenarios* 3), characteristic of all efforts to predict the future is an imperative to act. As Horn explains: "You either want to know what to do or you want to find out what to prevent" (3).[14]

So, also without any judgment on the likelihood of occurrence, climate change scenarios are a powerful means to strategically "futurize." As Simone Pulver and Stacy VanDeever explain, "Scenario analyses are commissioned with the stated goal of aiding decision-makers in envisioning, understanding, and planning for the future" (1). They thus always adhere to certain premises and power structures. Moreover, in their effort to understand an unprecedented and complex phenomenon such as climate change, the scenarios cannot necessarily be explained on the basis of the extrapolation of current trends, but are dependent on the parameters of a storyline that the modelers decided to begin with. By putting data into meaningful sequence and giving it a graphic framework, visineers ultimately produce not only images, but also narratives consistent with particular worldviews.[15] These rely on science, socio-economic trends *and* the imagination. Paul Raskin draws a parallel between the set-up of scientific scenarios and theater: "In the theater, a scenario is a summary of a play. Analogously, development scenarios are stories with a logical plot and narrative about how the future might play out" (14). Accordingly, the creation of a scientific image, in this case a climate change scenario, resonates a particular assumption about the course of development of the world. Thus no climate scenario is ever neutral or unambiguously true.

Science studies scholars have stressed this point more generally in their investigation of science and its practices, emphasizing that objective research results that represent "natural facts" are always also culturally constructed and determined by a certain network of power relationships. Most notably, Bruno Latour and Steve Woolgar's social constructivist reading in *Laboratory Life* questioned the self-acclaimed objectivity of science by arguing that ultimately all "truths" scientists arrive at are the results of social interactions and interpretations. According to them, "scientific activity is not 'about nature,' it is a fierce fight to *construct* reality. The *laboratory* is the workplace and the set of productive forces, which makes construction possible" (243).[16] Latour's work in general was groundbreaking because it questioned the vehemently proclaimed distinction between the hard sciences (natural sciences) and the soft sciences (humanities) by explaining that both disciplines are ultimately based on processes of interpretation and prioritization—something that natural scientists too often neglected by presenting their results as factual truths. Scientific knowledge, hence, is far from neutral, but fulfills a certain function: that of producing a certain kind of knowledge. Failing to reconcile the "two cultures," responses to Latour's argument as well as other constructivist positions climaxed in the "Science Wars" (Andrew Ross) of the 1990s and disciplinary reservations continue until today.[17]

Climate change notably challenges science as an "objective truth" and forces us to consider more closely what is real and what is constructed. However, in light of recent "deconstructivist" arguments made by climate change deniers, who highlight the lack of scientific certainty, Latour has commented on the thin line between "conspiracists'" and social scientists' critique, thereby reflecting on his own approach: "While we spent years trying to detect the real prejudices hidden behind the appearance of objective statements, do we now have to reveal the real objective and incontrovertible facts hidden behind the illusion of prejudices?" ("From Matters of Fact" 227). Pointing toward his doubts about having exposed the social construction of scientific facts, and wondering about the purpose of such a critique, Latour ultimately encourages critics to discuss "matters of concern" with particular caution and care (246).[18] Discussing issues like climate change in terms of their constructedness thus should not amount to denying their reality.

With this in mind, my work sets out to expose the hidden cultural narratives embedded in scientific climate change discourses without discrediting science or denying the material effects of climate change. Following

Latour's observation, "The critic is not the one who debunks, but the one who assembles. The critic is not the one who lifts the rugs from under the feet of the naïve believers, but the one who offers the participants arenas in which to gather" (ibid.). Pointing to the imaginative potential of climate change scenarios does not serve to cast doubt on climate science but instead highlights what so far has been neglected due to a dominance of discussions about its accuracy. In lieu of calling for more certainty of scientific data, new ways of engaging with climate science are needed—ways that allow for a "deeper level of complexity: the interrelated personal, aesthetic, social, and political choices people make as they react to changes in the climate around them" (Trexler, "The Climate Change Novel") and that open up the discussion of climate change to a wider public.

Without diminishing the materiality of climate change, the only means of perceiving climate risks is through discourse, narratives, climate models and visualizations.[19] While all are equally valid tools to think about the future, graphic scenarios have been the prime means of understanding and communicating climate change as a physical phenomenon for a long time. As argued above, this has led to a techno-managerial discourse about how to "fix" the climate, which fails to grasp the personal, emotional and more intimate dimensions of the multifaceted and individual as well as global crises that climate change harbors. Thus, instead of rooting the climate narrative in knowledge claims of natural sciences and framing it as a numerical problem that can be solved (Hulme, *Why We Disagree*), there is a need to understand "climate change as a political and social construction *as well as* a physical reality" (Skrimshire 2–3). In fact, and again turning to Latour, "matters of concern" like climate change work in-between the dualisms of modernity. They are "much more social, much more fabricated, much more collective than the 'hard' parts of nature ... On the other hand they are much more real, nonhuman and objective than those shapeless screens on which society—for unknown reasons—needed to be 'projected'" (Latour, *We Have Never Been Modern* 55). In order to understand global warming, then, we should not solely consider its physical qualities and scientific explanations but see it within a broader network of discourses and social practices. Mike Hulme has also prominently called for a re-examination of the debates around climate change, including contributions from the "interpretive humanities," science studies, as well as geography ("Four Meanings" 39).

Indeed, in the past years an increasing number of scholars from the humanities and social sciences have begun to engage with climate change in order to reveal the "social imaginaries" lying underneath any description of the phenomenon. As Bronislaw Szerszynski and John Urry explain, "Any description and prediction of climate change and its impacts is entangled with specific imaginaries of how society is, and how it ought to be" (4). Consequently, "Climate change is always already social; the social does not need to be added to it, just to be revealed" (ibid.). Climate change therefore figures as a proxy for a variety of political, economic as well as environmentalist projects such as the affirmation of sovereign power, the commodification of CO_2 in emissions trading, or the endorsement of vegetarianism (Hulme, *Why We Disagree*).[20] Accordingly, an awareness of the social and cultural narratives, which shape the very idea and interpretations of climate change, is essential to grasp the full complexity of climate change, and, in particular, "to attend more closely to what we really want to achieve for humanity" (Hulme, "Four Meanings" 55). Especially because the focus of debate is on the prospective dimension of climate change, but scientific papers only reach a very specific audience and remain problematic tools of engaging with the future, it becomes inevitable to rethink their authority over the representation of climate change.

The importance of literature for the discussions of climate change lies precisely in its potential to offer imaginaries for the unfamiliar realm of the future. It thereby not only reframes scientific data in a way that provides insight into the intimate aspects of human struggles in altered environments, exposes potential conflicts and is able to create affect, but shapes the very idea of climate change. Robert Macfarlane also underscores the significance of literature, which according to him can relate "hypothetical situations—alternative lives, or futures, or landscapes—as though they were real." Importantly, climate change fiction does not solely depict the future "as though it were real." In fact, it is not mere illustration but actively contributes to conceptions of climate change. Seconding this idea, cultural studies scholar Emily Potter foregrounds the importance of rethinking the role of representation in environmental discourse. Quoting her observations in this context, proves to be insightful:

> Text doesn't just mediate the world, but is active in its making. At the same time, of course, we know the world in more than textual ways … The demands of environmental change expose the limits of representation but they also inspire different ways of thinking about how we come to know. Through a revised approach to representation we can keep a host of realities in tension as we seek to find meaning. This won't mean that we know

> climate change and its impacts in any final way. It might mean, however, that we have registered its composing complexity, and taken a stake in what it will be. (Potter)

In line with Latour's discussion of matters of concern, Potter explains here that there is no single way in which one can know or communicate climate change. While such an understanding of climate change as an assemblage of realities and meanings constantly in flux may pose challenges to the politics of climate change because it does not easily comply with a utilitarian approach to knowledge, it opens up the space to acknowledge different ways of "knowing" climate change such as, for example, through fiction.[21] Breaking with "closed world views and exclusionary truth-claims in favor of plural perspectives, multiple meanings and dynamic interrelationships" (Zapf, "New Directions" 148), literature makes an important contribution to our understanding of climate change—one, which admits to the validity of a multitude of narratives that constitute the meaning of the phenomenon. This cultural repertoire of future imaginaries is crucial to reflect and critically re-assess what can be done to create more sustainable futures.

Moreover, literature is particularly well suited to address some of the representational challenges global warming poses. Climate change is a phenomenon that occurs on a global scale. Furthermore, its impact plays out on a range of timescales, with its most detrimental outcomes lying in the future. It thus calls for a renegotiation of short-term thinking and long-term environmental consequences. The complexity of the climate system as well as its feedback relation with human action, however, impedes a precise prediction of the ramifications of climate change. It thus produces a variety of uncertainties, which alienate people from taking responsibility. Another challenge is that global warming profoundly unsettles the dualistic modern thought predominant in Western societies, which is based on philosophical division of nature and culture. While on the one hand, it needs to be acknowledged that climate change is, indeed, a human-authored transformation, this cannot serve to validate a bifurcated human-nature view, which postulates that the environment is something "out there" upon which human actions are played out. Instead, human and non-human actors are co-creators of climate change—an understanding, which, nevertheless, does not absolve humans from taking responsibility toward environmentally hazardous activities. One of the greatest challenges of climate change that tends to get sidelined in climate scenarios and economic debates is precisely the question of justice and responsibility. Since the effects of global warming will be most severe for those who contribute the least to rising greenhouse gases, thinking about climate change also means thinking about the social inequalities it may aggravate.

For writers engaging with anthropogenic climate change, dealing with these challenging traits named above is paramount to narrating climate change. Indeed, with the turn of the millennium an increasing number of writers started to take on and address these challenges, producing a growing repertoire of works for literary analysis. Their books offer compelling stories, which entice a planetary thinking and a contemplation of different timescales, guide readers through future uncertainties, remind them of their moral responsibilities, and explore the world as a—what I would call—climateculture hybrid.[22] In providing vivid imaginaries that bear justice to the representational challenges, climate change fiction actively shapes how we come to know climate change.

An analysis of climate change fiction is not only important because it reveals a significant cultural knowledge of climate change, but also because the articulation of climate change's representational challenges has several implications for the formal dimension of these works. More importantly, climate change fiction serves to critically reassess some of the dominant tropes and ideas discussed in ecocriticism—the discipline that also forms the guiding school of thought for this book. Ecocritics interested in climate change fiction have to not only move beyond generic biases but also prevalent ideas of how to communicate environmental crises. Instead of apocalypse, the climate change fictions discussed here, for example, present a dystopian vision of the future that corresponds with contemporary Western society's "dwelling in crisis" (F. Buell, *From Apocalypse* 173).[23] Further explained by Frederick Buell in *From Apocalypse to Way of Life*, environmental crisis has become a normal feature of everyday life. Consequently, rupture and revelation no longer work as "wake-up tools" in a world, in which complex processes pose constant threats to society. Thus, "instead of being haunted by a sudden world-end … we more realistically worry about a world that doesn't end, but which descends and further descends" (Beck qtd. in F. Buell, "A Short History" 29). The future scenarios depicted in the climate change fictions discussed in this book also no longer rely on apocalyptic end time visions. Instead they attend to the "slow agent," climate change, and dramatize the planetary transformations in new and innovative ways.

The chapter "Climate Change Fictions in Context: Socio-Politics, Environmental Discourse and Literature" elaborates in greater detail on the formal challenges climate change poses for environmental discourse and, in particular, for ecocritical analysis. By means of a brief historical overview of environmental crisis discourse since the publication of Rachel Carson's *Silent Spring*, the chapter identifies those characteristics that make climate change a difficult issue to be communicated. Drawing on Beck's risk theory serves to further illustrate how climate change differs from other environmental problems. As the discussion in Chap. "Climate Change Fictions in Context: Socio-Politics, Environmental Discourse and Literature" finally shows, not one singular narrative template communicates best this deterritorialized, accumulative and long-term crisis. Climate change fiction proves to be insightful in this context because it gives way to the multiple and innovative frameworks of telling contemporary environmental crisis. Just as climate change inspires new imaginaries, it also prompts ecocritics analyzing them to re-think their approaches and (discipline) dominating concepts. Positioning climate change fiction within a short genealogy of ecocriticism in the final part of Chap. "Climate Change Fictions in Context: Socio-Politics, Environmental Discourse and Literature" serves to illustrate this more in depth. The discussion also juxtaposes climate change fiction with other literary works that engage with climate-related phenomena and thereby aims to arrive at a more narrow definition of this growing body of works. Furthermore, it elaborates on the criteria that define the works chosen for analysis as climate change fiction. While I did not give preference to a particular genre (science fiction, thriller or the novel), three criteria guided my selection process: first, their explicit engagement with anthropogenic climate change; second, their depiction of a future or near future climatically changed world; and third, an ability to qualitatively sustain a literary analysis. Even though also several recently published works of nonfiction on climate change include (science) fictional elements about a climatically changed future, they nevertheless remain restrained by generic conventions and thus often closely adhere to the narratives of science. However, centering on the imaginary dimension and explicitly on the potential of fictional texts to overcome some of the representational challenges of climate change, this book focuses on works of fiction. Moreover, while a comparative approach would certainly provide valuable insights especially with a global phenomenon such as climate change, this project focuses on American works of fiction. As explained in the beginning of the first chapter, climate change plays a highly controversial role in the United States. An overview of these "socio-political climates" demonstrates that power structures play an important role in defining risks as such and setting risk agendas. Looking at novels from one particular national context then allows exploring if and how these works contribute or contest these dominant national discourses and whether they still manage to address climate change as the global phenomenon it is.

In this book I analyze twelve American fictional texts from science fictions to short stories, from books by well-established authors such as T.C. Boyle to newcomers such as Steven Amsterdam, and explore how these works convey climate change, deal with its challenging characteristics, and with what narrative techniques they ultimately participate in its communication. The textual analyses are structured into five chapters (Chaps. "Scaling Climate Change: The Transformation of Place in Climate Change Fiction", "Reimagining Time in Climate Change Fiction", "Manufactured Uncertainty: Climate Risks in an Age of "Heightened Security"", "ClimateCultures in Kim Stanley Robinson's *Science in the Capital* Trilogy", "Representing the Underrepresented: Climate Justice and Future Responsibilities in Climate Change Fiction"), each of which explores one specific representational challenge and the "Conclusion: Climate Change Fiction and the Introduction of New Genres in Environmental Crisis Discourse". Undoubtedly, there exist numerous works dealing with climate change. However, the text selection and the individual integration into a chapter occurred on the basis of the text's representation of a specific feature of climate change. The works that were finally chosen are emblematic but not in an exclusionary way. As a result of the complexity of climate change, my work moves beyond a solely ecocritical analysis and touches on a wide range of issues and disciplines. This also translates into the structure of this book, which instead of setting a separate theory chapter before the analyses, provides close reading chapters informed by relevant theory.

The analyses begin in Chap. "Scaling Climate Change: The Transformation of Place in Climate Change Fiction" with a discussion of how Steven Amsterdam's *Things We Didn't See Coming* and Barbara Kingsolver's *Flight Behavior* communicate the global transformations that climate change involves. Both works portray the ways global warming challenges place attachments and locally based forms of belonging. They suggest that climate change calls for new ways of thinking about the human place on the planet. Drawing on globalization theory and Ursula Heise's idea of eco-cosmopolitanism, this chapter examines Kingsolver's depiction of global connectedness, on one hand, and Amsterdam's global uprootedness on the other. It illustrates that the two climate change fictions employ very different narrative strategies to foster a planetary understanding of the world, which is vital for action against climate change. Moreover and contrary to the prominence of place-attachment in much environmental discourse, Chap. "Scaling Climate Change: The Transformation of Place in Climate Change Fiction" shows that emphasis on locality no longer suffices to adequately address a global phenomenon such as climate change.

While spatial configurations provide the central focus in Chap. "Scaling Climate Change: The Transformation of Place in Climate Change Fiction", the subsequent chapter examines the representation of climate change's temporal dimensions. "Re-Imagining Time in Climate Change Fiction" analyzes how *A Friend of the Earth* and *The Ice Lovers* offer a "timescape" perspective by articulating the slow and long temporal dimension of climate change. T.C. Boyle and Jean McNeil both point to the dangers of solely focusing on the present-based, short-term thinking. While notions of "industrial time" (Adam), of time being measurable and controllable, dominate contemporary Western societies and often foreclose an awareness of underlying slow and invisible environmental temporal processes, the novels discussed in Chap. "Reimagining Time in Climate Change Fiction" foreground specifically the temporal dynamics of the climate crisis. Climate change fiction thus counters tendencies of "future discounting" and an immersion in the present by creating the narrative relationships between past, present and future, between human and environment, and rendering the otherwise invisible dynamics of this long-term environmental transformation visible. The chapter's analysis of the representations of time and temporal narrative techniques finally exposes the structures of "slow violence" (Nixon), an inherent part of climate change crisis, and draws attention to the state of apprehension that these texts foster, and which is, according to Rob Nixon, crucial in making readers witnesses to that which remains otherwise imperceptible (15).

As sociologist Ulrich Beck has argued, contemporary problems like climate change challenge societal preoccupation with the calculability of the future. Emerging from a complex system, such problems are uncontrollable and epitomize what Beck has called "manufactured uncertainty." This makes decisions on climate mitigation particularly difficult. Also participating in the communication of risks, authors like Nathaniel Rich acknowledge that the future is open and a diverse range of scenarios probable. Yet, unlike scientific scenarios, novels usually do not fall victim to the claims of facticity as described in the foregoing discussion.[24] Instead, their imaginary realm allows for a reflection on climate change without being expected to provide certainty or solutions. They therefore present creative ways to respond to varying uncertainties in everyday life. Moreover, as shown in Chap. "Manufactured Uncertainty: Climate Risks in an Age of "Heightened Security"", Rich's *Odds Against Tomorrow* critically questions the "commodification of contingency" (Lobo-Guerrero), and exposes the power and profit-making structures at play in defining risks as such. It also addresses the dangers of fear mongering communication of climate risks. Nathaniel Rich thereby guides readers through a complex network of uncertainties while emphasizing that these uncertainties do not justify denial or neglect of climate change.

Engaging in yet another challenge that climate change poses, Chap. "ClimateCultures in Kim Stanley Robinson's *Science in the Capital* Trilogy" takes a closer look at Kim Stanley Robinson's trilogy, *Forty Signs of Rain*, *Fifty Degrees Below*, and *Sixty Days and Counting*. It focuses on the discussion of how a reciprocal relationship of nature and culture can be narrated in a way that acknowledges climate change as a hybrid entity—as cosmopolitan (Hulme), or what I would call a *climateculture* product. Part of the uncertainty discussed in the previous chapter derives, indeed, from the question of the extent of humans' impact on the global climate system. This, again, assumes a dualistic perspective: humans are acting *on* the environment. In contrast, as Robinson's trilogy reveals the intricate network of human and non-human relations, it also explores the mutual conditioning of the supposedly separate realms. It thus proves to be an interesting example of climate change fiction that goes beyond nature-culture binaries. In fact, it exemplifies what Hubert Zapf has termed a "reintegrative interdiscourse," staging how "the dichotomies of mind and body, intellect and emotion, culture and nature are overcome, and different areas of knowledge … are brought together in ecosemiotic networks of signs that transgress the separation between disciplines and cultures" ("Literary Ecology and the Ethics of Texts" 13). While acknowledging the complex dynamics that shape *climatecultures*, Robinson, nevertheless, emphasizes humans' moral responsibility to sustainably inhabit, or in the case of the trilogy "terraform," the planet.

In the final chapter, the question of responsibility is also of importance for the discussion of Octavia Butler's *Parable of the Sower*, Dana Stein's *Fire in the Wind*, Paolo Bacigalupi's "The Tamarisk Hunter" and

Benh Zeitlin's *Beasts of the Southern Wild*. More specifically, these works focus on issues commonly underrepresented in scientific discourse and "excommunicated" from the public sphere, but crucial to the climate change debate: they highlight that thinking about climate change and planning for more sustainable futures also requires more careful consideration of those people who are most vulnerable to global warming. These works complicate the idea of one collective climate change narrative and instead remind readers of the disproportionate effects climatic transformations have. Environmental justice scholars have pointed to the correlation between environmental and social problems and examined how poor and marginalized communities are especially vulnerable to environmental disasters. The climate change fictions discussed in Chap. "Representing the Underrepresented: Climate Justice and Future Responsibilities in Climate Change Fiction" also expose these environmental injustices, but simultaneously articulate a variety of justice claims that envision possible adaptation plans that could counter them. Climate change fiction thereby prompts questions concerning the ethical dimension of global warming.

The conclusion returns to the idea that every vision of the future also involves an implicit summons to take action. By comparing the climate change fictions, the conclusion sheds light on their underlying social projections, which is of importance especially in relation to the national context in which these narratives emerge. It provides insight into what sociopolitical tendencies are criticized and which ones endorsed. Consequently, the climate change fictions are discussed with respect to their magnifying and mirroring function. Combining these ideas, Mike Hulme has argued that climate change serves as a magnifier because it focuses "our attention on the long-term implications of short-term choices" and a mirror because it helps "to attend more closely to what we really want to achieve for ourselves and for humanity" (*Why We Disagree* xxiii). The climate change fictions discussed here clearly call for more social justice and sustainable living practices. Finally, the last pages also offer concluding thoughts on the characteristics in content and form of climate change fiction as an emergent new genre. It particularly focuses on the shortcomings of apocalypse in narrating climate change.

In its entirety, this book aims to make a contribution to the growing body of scholarly works concerned with climate change in the environmental humanities and more specifically in ecocriticism. Drawing on risk theory, globalization theory, critical science studies as well as literary and cultural theory, it explores the complexity of climate change and enriches ecocritical discussion by opening new pathways of inquiry. In taking a closer look at how novels deal with the representational and conceptual challenges that climate change poses, this analysis specifically engages in a critical discussion of the ideas at the core of ecocritical practice such as place, nature and justice and introduces concepts such as time and risks as equally important aspects. As writers come up with innovative narrative means to overcome the elusiveness of climate change, their works also constitute an insightful cultural repertoire valuable for discussion in the environmental humanities in general. While climate change fiction may provide a "depragmatized" knowledge, I would still agree with Kathryn Yusoff's claim that "aesthetics must be considered as part of the practice of politics; a space where things are made, both materially and semiotically (to paraphrase Haraway) and a space that configures the realm of what is possible in that politics" (77). In this respect, climate change fictions play an important role in shaping not only our understanding of this unprecedented crisis, but also in guiding our responses to and actions against it.

Notes

1. All headlines are taken from the November 12, 2008 *The New York Times* issue distributed by The Yes Men.
2. Culture jamming is usually defined as a form of "guerilla" communication. As Mark Dery explains "culture jamming … is directed against an ever more intrusive, instrumental techno culture whose operant mode is the manufacture of consent through the manipulation of symbols" ("Culture Jamming"). It uses the same medium of communication that it attempts to criticize and often disguises its messages through irony or satire.
3. Certainly, the choice of headline concurred with the tabloid style of the newspaper the Yes Men used as a template.
4. Ursula Heise has pointed out that in contrast to the "poetics of authenticity," which appeals to a fixed eternal ecological state which needs to be preserved, "The poetics of responsibility recognises that every inflection of Earth is our inflection, every standard our standard, and we should not disguise political decisions about the kind of world we want in either the discredited objectivity of natural order or the subjective mystification of spiritual intuition" ("Teaching Ecocritical Theory" 53, sic).

5. With the exception of Ben Zeitlin's film *Beasts of the Southern Wild*, which I analyze in Chap. "Representing the Underrepresented: Climate Justice and Future Responsibilities in Climate Change Fiction", I focused on literature for this book. Interestingly, the Yes Men also draw on the "written word" for their cultural jamming actions on climate change, which seems best suited with its imaginative potential to address some of the challenges that climate change poses.
6. I am using the terms climate change and global warming interchangeably in order to avoid repetition. However, I am aware of the discussions evolving around the term "global warming" for a phenomenon that will not only lead to an increase in temperature but also to numerous other climatic changes.
7. While most of the works discussed here were written by US American authors, there is one climate change novel, *The Ice Lovers*, whose author is Canadian. I have nevertheless included Jean McNeil's work for analysis, because it serves as an interesting example for the portrayal of slow environmental time. Moreover, for a long time the socio-political context and position toward climate change has been similar in the US and in Canada.
8. The paradox here is, of course, that also authors of climate change fiction rely on scientific accounts. While some draw more closely on scientific reports than others, all of these works are relational to climate science and in affirming its projections they also give rise to its legitimacy. This is discussed more in depth in Chap. "Manufactured Uncertainty: Climate Risks in an Age of "Heightened Security"".
9. In 1959 the physicist C.P. Snow, most prominently, pointed out that in Western intellectual life he had noticed a widening gulf between "two cultures"—one represented and constituted by natural scientists and the other by writers and humanities scholars. Snow, himself, credits the natural sciences with rather positive attributes important for progress while the humanities remain an old-fashioned discipline mainly interested in the past to him. He, nevertheless, argues that in trying to find solutions to problems in increasingly complex societies, ethical and moral questions also had to be considered and thus the gulf of misunderstanding between the disciplines needed to be overcome (Snow 98). Snow's claim of the existence of the two cultures did not only provoke a controversial debate at the time but continues to influence contemporary disciplinary practices and discussions.

10. As Bruno Latour explains in *We Have Never Been Modern*, the division between science and *lettre*, nature and culture is a byproduct of modernity and the scientific revolution in the seventeenth century (29). According to him the "modern constitution" is characterized by the belief that there is no relation between nature and culture and the disciplines that address them. The Cartesian division between body and mind, which formed our perception of nature as something innate that should be dominated, epitomizes this idea.
11. As Gereon Uerz in his comprehensive historical overview of the meaning of future envisioning for the perception and interpretation of the present reality of a certain society in time, points out that no matter the method or content, future envisioning is always a hybrid of fiction and reality that can never be marked as fact or fiction, right or wrong at its point of articulation (13).
12. The Special Report on Global Emissions Scenarios (SRES) published in 2000 represents the most prominent future simulations in climate change research. They consist of 40 different scenarios projecting future greenhouse gas emissions based on four different underlying "*narrative* storylines" (driving forces). The storylines vary according to economic development, technical advance, land use, political situation (global or local actions), population growth, etc. They were developed by an interdisciplinary and international writing team, which includes scientists as well as NGO representatives. While, for example, in an A2 scenario, which takes as premise a very heterogeneous world with high population growth, slow economic development and slow technological change, global CO_2 emissions almost exponentially rise, they steadily decrease in a B1 scenario, which represents a sustainable development (IPCC, *Emission Scenarios* 3–8).
13. Horn draws on Herman Kahn's title "Thinking about the unthinkable" in order to elucidate how scenarios allow for thinking about phenomena that are experientially unprecedented and cannot be experimentally explored. For the differentiation between prediction, prognosis, projection and a more elaborate analysis of the scientific functioning of scenarios see Gramelsberger's article on the potential of projection of climate models.
14. My own translation from the original: "Entweder man will wissen, was man tun soll, oder herausfinden, was verhindert werden muss" ("Der Anfang vom Ende" 3).

15. Visineers is the commonly used term for the designers of visualizations such as, for example, the emission graphs (Schneider 86).
16. Latour later distanced himself from poststructuralism and a radical constructivist approach. For further insights into constructivist approaches in science studies at the time see also Thomas Kuhn.
17. In the 1980s an increasing number of publications engaged with postmodernist and poststructuralist ideas powerfully building on the idea that every reference to some natural essence was a linguistic and cultural construct (see Zapf "The State of Ecocriticism"). At the same time, the humanities and social sciences also turned to a critical investigation of science and its practices. When the physicist Alan Sokal managed to publish an article on quantum physics in *Social Text* (1996), he shortly afterwards disclosed that this had been an academic hoax. According to Sokal his article purposely included nonsense in order to demonstrate the arbitrariness of the then prominent social constructivism and postmodernism. This incident catalyzed the "Science Wars," which were further spurred by the Western press (see Heise "Cultures of Risk"), and eventually broadened the gap between the humanities and natural sciences.
18. In *Politics of Nature*, Latour elaborates more in detail on the distinction between "matters of fact" and "matters of concern" (also clarifying in a footnote that with "matters of concern" he is re-naming his earlier employed term, the "quasi-object," discussed in *We Have Never Been Modern*). In contrast to earlier times, when objects were considered "matters of fact," that is risk-free, clearly defined, with well recognized properties (22), and thus easily upholding the illusion of objective science, Latour sees a current proliferation of "matters of concern." Matters of concern have numerous entanglements, no singular and clearly determinable producer, and are very difficult to be "naturalized": "To deal with them, we do not have the social or political world on one side and the world of objectivity and profitability on the other" (Latour, *Politics of Nature* 24). Matters of concern, rhizomatic and incommensurable, thus call dualistic and hierarchical thinking into question.
19. The future scenarios of climate change discussed here are simultaneously and inevitably risk scenarios. Chapters "Scaling Climate Change: The Transformation of Place in Climate Change Fiction" and "Manufactured Uncertainty: Climate Risks in an Age of "Heightened

Security"" elaborate more in detail on risk theory and the importance of literature for the communication of risks.

20. The special edition "Changing Climates" of *Theory, Culture & Society* (2010) also provides further insights into the social and political narratives of climate change.
21. Also Hubert Zapf explains that literature produces a depragmatized kind of cultural knowledge—one, which cannot be easily translated into politics and action, but which accounts for the interrelated ethical and ecological dimensions that other forms of knowledge neglect ("Literary Ecology and the Ethics of Texts" 13).
22. See also my discussion in Chap. "ClimateCultures in Kim Stanley Robinson's *Science in the Capital* Trilogy". The idea of climateculture draws on Donna Haraway's concept of natureculture (*Simians, Cyborgs*), which centers on an understanding of "the human as part of an assemblage of what is reality" (Potter) rather than the origin of meaning-making. Nature and culture are, accordingly, always co-constitutive.
23. More generally, of the over 200 novels reviewed for his climate change fiction survey, Adam Trexler points out, that there is only one, *The Earth Party* by George Marshall, which describes a positive scenario, a successful ecological revolution, related to climate change.
24. Interestingly enough, though, also literary or fictional works have been criticized for their inaccuracy in representing the facts, the most prominent example being the movie *The Day After Tomorrow* as also Leiserowitz has argued.

Climate Change Fictions in Context: Socio-Politics, Environmental Discourse and Literature

Socio-Political Climates

Everybody talks about the weather, but nobody does anything about it. (Mark Twain, 1897)

The American way of life is not open to negotiation. (President George H.W. Bush at the "Earth Summit" in Rio de Janeiro in 1992)

In his 2013 inaugural address, President Barack Obama stated, "We will respond to the threat of climate change, knowing that the failure to do so would betray our children and future generations" ("Inaugural Address by President Barack Obama | The White House"), thereby announcing a change of direction in US politics and prominently affirming the overwhelming evidence by climate scientists of the reality of global warming. Climate change, according to the United Nations Framework Convention, describes a "change of climate which is attributed directly or indirectly to human activity that alters the composition of the global atmosphere and which is in addition to natural climate variability observed over comparable time periods" (United Nations, "Article 1: Definitions"). Due to the increase of greenhouse gas (GHG) emissions since preindustrial times, atmosphere and ocean temperatures have warmed significantly, polar ice has melted, and sea levels have risen (IPCC, *Climate Change 2013* 2).[1] The main atmospheric GHGs that critically endanger the balance between input from energy of the sun and loss of that energy into space

A. Mehnert, *Climate Change Fictions*,
DOI 10.1007/978-3-319-40337-3_2

(i.e., the planetary energy budget) are carbon dioxide (emitted in burning fossil fuels), methane, and nitrous oxide (emitted in modern agricultural production) (Boykoff 2011, Lever-Tracy). According to a 2013 report released by the Intergovernmental Panel on Climate Change (IPCC), global warming is likely to continue, and even if emissions are significantly reduced, many aspects of climate change will persist for centuries (25).[2] Impacts of future climate change on ecosystems (biodiversity loss), resources (food and water crises), and on settlements and society (displacement) are projected to be severe.

When talking about climate change, the United States occupies a central position, not only because it is annually the second greatest producer of CO_2 emissions (until 2009 it was the first) through its domestic and international industries, but also because the climate change debate has experienced an unprecedented fervor and a high level of controversy in the US. Furthermore, though it consumes 25 percent of global energy annually, the US has so far been a laggard concerning federal or even international climate change policy and commitments (Giddens, *The Politics of Climate Change* 87). Yet, on a municipal and state (as well as recently on a federal) level, there has been significant progress in climate policies over the past years, and the variety and strength of environmental groups in the United States and their push for action on climate change is noteworthy. Since this publication focuses on US climate change fiction, and presuming that literature responds to and participates in the socio-political context in which it is produced, a short overview of American political trends regarding climate-related questions is necessary.

In the late 1950s, Professor Roger Revelle and a research team around David Keeling of the Scripps Institute for Oceanography started to measure daily CO_2 concentrations in the Earth's atmosphere. After a couple of years, their findings showed a steady increase in CO_2 concentration in the air, and scientists began drawing some initial alarming conclusions regarding the connection between increased CO_2 concentration and increased temperatures.[3] As a growing number of scientists became concerned with the warming impact of the increase in greenhouse gases, they also pushed for the recognition of the issue within the US public and political realm (Lever-Tracy 4). However, it wasn't until 1988 that the issue reached prominence. As Dessler and Parson describe, that year the US experienced an extreme heat wave and a subsequent drought period, which facilitated a general awareness of the earth's fragility and led to first debates about climate change (22).[4]

The United Nations Environment Program and the World Meteorological Organization established the IPCC (in 1988) to advance climate science, representing the first major international response to the growing concern about climate change. In 1992, international governmental representatives joined together at the "Earth Summit" in Rio de Janeiro and reached the first international treaty on climate change, the Framework Convention on Climate Change (FCCC). Following the "common but differentiated responsibility" principle, the treaty determined the "stabilization of greenhouse gas concentrations in the atmosphere at a level that would prevent dangerous anthropogenic interference with the climate system" (United Nations, "Article 2: Objective").[5] Even though the United States signed and ratified the treaty, it was among the many nations that only committed on a voluntary basis to relatively weak and preliminary measures in the treaty (Dessler and Parson 25). The Kyoto Protocol, proposed in 1997, was introduced to serve as a more legally binding framework.[6]

While the Clinton administration had participated in the negotiations of the Kyoto Protocol and eventually even signed it, the US Senate passed the Byrd-Hagel Resolution at the same time. This resolution opposed the signing of any treaty that would seriously harm the US economy and required developing countries to make specific commitments in accordance with the commitments being made by developed nations (Leggett 2). As a consequence, the Kyoto Protocol was never submitted to the Senate for ratification neither by Clinton nor by his successor. When George W. Bush became president of the United States in 2001, it became even less likely that the administration would push for the Senate to ratify an agreement to Kyoto: Questioning the accuracy of climatic research and highly influenced by the oil lobby, the newly elected president made it clear during his presidency that the United States would not adhere to any international agreement with binding CO_2 emission targets. Since President Bush argued against mandatory GHGs regulations, federal policies on emission-reducing technologies were only pursued halfheartedly and the country's absolute emissions levels increased (Selin and VanDeveer 122).

With the election of President Barack Obama, expectations rose again for a renewed effort in international climate change negotiations, especially since many signatories of the Kyoto Protocol were struggling to meet their targets and delegates to a global climate summit in Bali in 2007 had failed to negotiate new binding goals. Yet, it was only at the Conference of the Parties (COP) in Paris in 2015 that an international agreement was reached and still the question remains if enough member states (including the United States) will ratify the treaty. At the national level, President Obama's first call for far-reaching climate legislation grew quieter as the American Clean Energy and Security Act of 2009, which he had supported, was defeated in the Senate (Giddens, *The Politics of Climate Change* 88). However, he has now toward the end of his second legislation put forward a new policy, The Clean Power Plan, which aims to support his overall goal to reduce greenhouse gas emissions by 17 percent below 2005 levels by 2020.[7] However, with upcoming elections and great resistance by several states, the success of the plan remains uncertain. Anthony Giddens summarizes three reasons why the implementation of climate change policies in the Unites States has so far been so difficult: "(1) the President's dependence to negotiate decisions with Congress, (2) the great influence of lobbies on members of Congress, (3) the political polarization of climate change" (*The Politics of Climate Change* 89).

Though the United States has widely been considered a laggard in international and federal policy action on climate change, the failure to institute climate change legislation on a federal level provided an opportunity for experiments and positive developments in state policies. As Henrik Selin and Stacy VanDeveer elaborate, half of all US states have established GHG emission targets, and there have been joint efforts for cap-and-trade schemes such as the Regional Greenhouse Gas Initiative (RGGI) launched in 2009. Many states have implemented renewable portfolio standards and expanded on energy efficiency programs. Cities also participate in the climate change policy process, establishing their own emission reduction targets and engaging in the Cities for Climate Protection program (CCP) by the International Council for Local Environmental Initiatives (ICLEI) or the C40 network. As Selin and VanDeveer explain, this is far from an unusual development:

> Many of these actions are consistent with a long-standing US tradition of environmental federalism, where policy-leading subnational entities are the first ones to act on environmental issues such as air pollution, hazardous substances, and waste management, pushing federal policymakers and authorities to eventually set national standards and regulations. At the same time, many political leaders in states oppose far-reaching GHG controls, echoing political and economic reasons voiced by skeptics in Congress. (Selin and VanDeveer 123)

Despite all of these efforts, US national GHG emissions are likely to increase, as global warming remains a highly contested and divisive issue among the US population.[8]

US neglect of climate change policy cannot be explained with the argument that the country is an "unaffected bystander," since scientific calculations show that the United States will be significantly affected by the consequences of global warming. Extreme weather events, such as droughts, heat waves, heavy precipitation, and hurricanes are likely to increase, and sea levels are rising, threatening coastal communities (IPCC, *Climate Change 2007: Synthesis Report* 52). Furthermore, the melting of the arctic ice cap not only poses serious nutrition problems to the Inuit populations in Alaska, but may also ignite geopolitical conflicts over newly accessible resources in that area (Ingimundarson). While many of these effects will only be felt at a later point in time, average surface temperatures in the United States have already risen by 1.4 degrees Fahrenheit over the past century, and the record for the warmest year recorded is continuously broken (C. C. D. US EPA, "U.S. and Global Temperature"). Accounts of floods, droughts, and casualties from heat weaves figure prominently in the media. Still, "many Americans, especially those who self-identify as Republicans, ... do not view climate change as a problem that requires fundamental societal change or the introduction of policy measure that increase carbon or energy costs" (Selin and VanDeveer 126). As Giddens' analysis has already suggested, political and socioeconomic factors are not the only reasons for these views of climate change, as socio-cultural patterns of interpretation and perception also play an important role.

Opposition to climate change policy and mandatory GHG reductions is fueled by conservative and neoliberal think tanks. When global warming first became an issue of public interest in the United States, these think tanks mainly aimed at discrediting climate science, and powerful oil lobbies pushed to stress factors of scientific uncertainty over several years. The discrediting of Michael Mann's "Hockey Stick" graph, of the Congressional hearing statements made by James Hansen (head of the NASA Goddard Institute), and the manipulation of documents issued by the White House (as depicted in Al Gore's *An Inconvenient Truth*) are only some of the many examples of climate skeptics' techniques that influenced public opinion (Lever-Tracy 8–11).[9] As long as skeptics could call for more research, no action to curb greenhouse gas emissions needed to be taken. However, as the upward trend in global warming has become undisputable, confirming climate science models that indicate the correlation between greenhouse gases emission and rising temperature, the climate skeptics have shifted their focus toward decisions regarding actions taken against climate change (Boykoff and Boykoff). For example, they now argue that reducing GHG emissions will not only significantly increase expenses for households and firms alike, but also that reductions imply restriction in mobility and consumption—thereby challenging core American values. A Competitive Enterprise Institute slogan against control of CO_2 emissions ("They call it pollution; we call it life") illustrates the ways in which climate change has been read as questioning the foundation of American lifestyle (Selin and VanDeveer 126).

Climate change is thus framed as a serious threat to the American way of life. In particular, the ability to move and expand seems to be at stake. Mobility has historically played an important role with settler culture and the idea of "open frontiers" in the United States. There always existed the possibility of moving westward if resources or land were no longer sufficient. Even when Frederic Jackson Turner declared the frontier to be officially "closed," in his famous 1893 essay, Americans continued searching for new open spaces "in outer space and cyberspace and even below the ocean's surface [and also in the atmosphere as recent ideas on geo-engineering could be interpreted as conquering the last frontier]. The frontier remains a potent symbol more than a century after it physically disappeared" (Mintz). This idea of mobility—epitomized in car ownership—still prominently figures in everyday life in the United States. As framed by conservative lobbies, emission reduction would thus imply profound changes not only for the tens of millions of commuting workers, but also for Americans in general who drive to the supermarket, the mall, their friends' houses, etc. It may thus seem an easier choice for many to deny the existence of climate change even if confronted with extreme weather events than to change one's lifestyle, especially if one's living environment and infrastructure is not well-prepared for such a change.

Further skepticism toward climate change comes from the Religious Right in the United States, which uses both radically conservative as well as nihilist arguments in opposing policy changes. Leaders of the Religious Right explain that climate change couldn't possibly be real because God wouldn't allow humankind to mess with his Creation, and, on the other hand, even if it is real and destructive, that radical change in its "creative destructiveness" is likely to be a good thing. As recent Republican presidential candidate (and Religious Right favorite) Rick Santorum stated, "We were put on this Earth as creatures of God to have dominion over the Earth, to use it wisely and steward it wisely, but for our benefit not for the Earth's benefit" (in Stafford, n.p.). Furthermore, dispensationalist tendencies among religious groups have proven to be counterproductive for environmental policy (Wardekker, Petersen, and van der Sluijs

55). Discarding concern for the future of the planet, some fundamentalist Christians welcome destruction and ultimately apocalypse, since, according to the doctrine of the Rapture articulated in the Book of Revelation, they are expecting salvation in heaven while sinners will be left on earth.[10] Other religious groups have emphasized that it is not in human hands to alter the climate or interfere with the providential path.

Climate change hence remains a controversial issue in the United States. Reasons for climate change skepticism and denials are so manifold and conceptualization of global warming so diverse that it seems too simplistic to subsume climate change under the meteorological master narrative of "humans increase greenhouse gases" (Forsyth 21). Instead, the multifacetted socio-political and cultural aspects of the issue must be considered in explaining this otherwise elusive issue. Moreover, it is wise to consider the perspective of Ulrich Beck, who maintains that "hazards are subject to historico-cultural perceptions and assessments which vary from country to country, from group to group, from one period to another" (Beck, *Ecological Politics in an Age of Risk* 91). Elaborating on how climate risks are shaped, contested, and understood within a national context can thus provide valuable insights even when discussing a global phenomenon such as climate change. As illustrated above, the power over discourses plays an important role in defining climate risks as such. Since right-wing lobbies successfully downplayed the risks of global warming for a long time, climate change as climate crisis has only emerged in the public sphere fairly recently. Even then its reconfiguration as apocalypse led to further polarization and the alienation of the environmentalists' cause. As the following discussion of "environmental crisis discourse" shows, climate change—in comparison to earlier environmental problems—poses communicational challenges that may require environmentalists and others to go beyond the often-employed apocalyptic rhetoric. It raises questions about what kinds of narrative frameworks would best communicate the urgency of combating this particular phenomenon.

Climate Change in Environmental Crisis Discourse

When climate change appeared on the public agenda in the mid-1980s, it joined a variety of environmental problems that had become sources of increasing public concern since World War II. While earlier conservationists approaches of environmentalism focused on the creation and preservation of national parks and of natural resources, the modern environmental movement revolved around efforts to address growing pollution and environmental hazards. The rapid suburban sprawl of the postwar years had led to difficulties in maintaining sewage systems, as mass production for a growing population demanded more resource exploitation and simultaneously produced more waste. In addition, an ever-expanding automobile industry connecting cities and suburbia gradually worsened air pollution (Flippen 273–276). So just as a new leisure culture was forming, which granted Americans more time to appreciate nature, they were also becoming more conscious of its vulnerability. It was, without doubt, the publication of Rachel Carson's *Silent Spring* (1962) about the pesticide DDT that most prominently called attention to the growing detrimental environmental consequences of human activities and forcefully brought about the "disenchantment from the illusion of the green oasis" (L. Buell, "Toxic Discourse" 648).[11]

Carson's book has powerfully shaped environmentalism to the present day. At the time of the work's publication, however, its unique style (a mixture of popularized scientific writing and fable) as well as its content drew a lot of criticism from her scientific peers. Yet her narrative not only allowed for easy accessibility and therefore a wide readership, but also included the public in a debate that went beyond the usually enclosed scientific circles—a democratization of scientific research that Carson, a scientist herself, had intentionally sought. The environmental crisis thereby made front-page news and became omnipresent. As Frederic Buell explains, *Silent Spring* "put several aspects of the environmental crisis—toxification of ecosystems and human bodies—in terms so unforgettable that the force of her presentation was never equaled" (179). Carson had made it clear that the environment and public health were inseparable and that now both were endangered by the impacts of pesticides. Furthermore, Carson exposed for the first time very overtly that the growing environmental crisis was man-made and that humanity might be embarking on a suicidal mission, at the end of which loomed a "silenced nature." Analogous to the threat imposed by nuclear science, which resonated deeply in the public consciousness during the atomic arms race of the Cold War, Carson argued that agricultural research had created an equally dangerous synthetic weapon that was directed not only toward insects but foremost against ourselves (Killingsworth and Palmer 28–9). To give emphasis to her toxic discourse she draws on the trope of apocalypse.[12] In her prologue, a pastoral-seeming scene of a town is betrayed because "an evil spell had settled on the community" (Carson 2) and turned it into a poisoned and deadly place. Notably, and playing on the threat of nuclear destruction, the villains were "the people [who] had done it themselves" (Carson 3). Juxtaposing the initial peacefulness of the scenery to the dying cattle, fish, and birds, to the spreading of diseases and finally to the complete silencing of nature, Carson depicts an end of the world scenario, prompting readers to "change or die" (Killingsworth and Palmer 31).

Ever since Carson's work, apocalypse has figured as a powerful trope in fiction and nonfiction works depicting environmental problems, including toxicity, resource shortages, pollution, and other manmade hazards.[13] As Lawrence Buell tellingly remarks, "Apocalypse is the single most powerful master metaphor that the contemporary environmental imagination has at its disposal" (L. Buell, *The Environmental Imagination* 285). Indeed, books such as Paul Ehrlich's *The Population Bomb* (1968), published only a few years after *Silent Spring*, or the watershed publication *The Limits to Growth* (1972) deploy eschatological rhetoric to persuade readers of the dangers of overpopulation or resource depletion. Yet, Ehrlich's neo-Malthusian scenarios were not only problematic because he radically questioned libertarian ideas of individual autonomy, which further alienated environmental activists from the rest of society, but also because they ultimately did not materialize (Killingsworth and Palmer 33). Garrard thus concludes: "Ehrlich's strategic hyperbole might be justified in the interest of successful persuasion, but the long-term dangers this approach poses for environmentalist causes may outweigh its rhetorical usefulness" (*Ecocriticism* 99). Though Ehrlich himself cautioned that his scenarios were only possibilities, critics correctly pointed out that many only considered the accuracy of the science presented in his work and thus failed to recognize the political value of them.

The publication *The Limits to Growth*, though very popular, also caused great controversy. Commissioned by the Club of Rome, a group of scientists around Donella and Dennis Meadows modeled the development of the world based on the variables of population growth, resource depletion, pollution, agricultural usage, and economic trends. The publication was unique because it illustrated the systemic, complex, and dynamic character of environmental crisis and stressed its global nature (F. Buell, "A Short History of Environmental Apocalypse" 15). Apocalypse served here to stress the revelational function of the book, disclosing the limits of human existence on earth. It was also employed in this work to persuade readers to take action against the approaching environmental crisis. Along with these prominent nonfiction texts, environmental apocalypse also emerged in works of popular fiction at the time, such as Richard Fleischer's movie *Soylent Green* (1973) and books such as John Brunner's *The Stand on Zanzibar* (1968) and *The Sheep Look Up* (1972) that took up issues of overpopulation, scarce resources, and increasing pollution.

While eschatological rhetoric for a long time and up till today often seemingly serves as the most adequate discursive strategy to communicate the urgency of action needed against ever accumulating environmental hazards, its employment is also highly problematic. As Garrard explains, apocalyptic rhetoric often forecloses the possibility of compromise because of its tendency to polarize responses and conflate a multitude of problems under a general doom scenario of "environmental crisis" (*Ecocriticism* 105–107). Furthermore, apocalyptic rhetoric often feeds into and possibly even "produces" the crisis it attempts to describe (ibid.).[14] Erik Swyngedouw is even more critical of apocalyptic imaginaries because they, according to him, powerfully displace political and social debates and are thus "decidedly populist" (Swyngedouw 219). Moreover, he adds, environmental apocalypse no longer provides redemption but solely feeds on the fascination with ecological disasters and negative desires (ibid.).

While attention to the environmental crisis experienced a decline in public discourse during the 1980s and early 1990s, there was no significant halt in hazardous activities during those years and environmental problems did not decrease. On the contrary, "index after index showed biospheric conditions not getting better, but becoming much, much worse" (F. Buell, "A Short History of Environmental Apocalypse" 25). With the growing number of environmental disasters—especially well-mediatized man-made catastrophes such as the Bhopal incident (1984), Chernobyl (1986), the ozone layer depletion, the Deep Water Horizon oil spill (2010), as well as possible climate change–related events such as Hurricane Katrina (2005)—the devastating effects of human activities on the environment could no longer be neglected.

For more than two decades now, these as well as many other large-scale disasters have provided the basis for numerous environmental crisis publications.[15] Scholars and journalists have also turned their attention toward the presence of various toxic waste sites, ecological disasters due to ever more aggressive resource extraction, the loss of biodiversity due to forest clearances, and a multitude of other environmental problems. And though apocalypse continues to play an important role as a rhetorical device, Frederick Buell convincingly argues that apocalypse is also being reinvented. For one, he sees its diversification into "more intimately realistic portrayals of damage done and in discourses of human-environmental mourning" and secondly into "depicting … [humans'] conscious immersion in uncertainty and rising risk" (30). With the realization that the return to an undamaged state of the world has become impossible, apocalypse, according to Buell is turned into "a way of life" (F. Buell). Accordingly, "people [are] now, in short, *beyond* the limits and dwelling *in* crisis, not just being driven towards that point" (F. Buell, *From Apocalypse to Way of Life* 188).

However, this "domestication of crisis" (F. Buell, *From Apocalypse* 186) was brought about not only by an accumulation of catastrophic events, but also due to a change in the nature of environmental problems. As technological advancement allowed for increasingly complex systems (industrial, informational, transportation, etc.), it also became more difficult to control the processes it generated.[16] Whereas sources for former environmental problems seemed to be more easily locatable, now the complexity of techno-economic systems prevented layperson and expert alike from understanding how potential risks could be avoided. Moreover, modernization and globalization have facilitated processes of "disembedding" (Giddens, *Consequences of Modernity*)—that is, mechanisms that detach social, cultural, economic, or governing structures from their local contexts. Giddens points to the positive effects of these disembedding mechanisms (for example for knowledge production), while also warning that they may provoke novel risks as "resources or services are no longer under local control" and, more importantly, "there is a risk that the mechanism as a whole can falter, thus affecting everyone who characteristically makes use of it" (126). There is thus a paradoxical underlying structure at play: While complex systems may provide more security (because they draw on expert knowledge), they are also often based on the trust that we put in the functioning of these systems. The risk scenarios, however, that these disembedding mechanisms simultaneously produce threaten this basis of trust. One consequence is the growing unease of living in great dependence on systems, which the layperson can neither understand nor influence—a feeling, which amplifies the societal sense of dwelling in crisis described by Frederic Buell.

According to sociologist Ulrich Beck, these fundamental changes have transformed modern industrial society into a risk society. This risk society finds itself endangered by the self-inflicted hazards of its own modernization—a development that Beck and others call reflexive modernization. Whereas former (traditional) modernization meant the change in social traditions based on a complete revision thereof (i.e., from agricultural to industrial society), now modernization is undermining its own premises as an industrial society (Beck, *Risk Society* 14). As Beck explains, in reflexive modernization "the subject of this creative destruction is not the revolution, not the crisis, but the victory of Western modernization" (Beck, Giddens, and Lash 2). In that sense, new modernity has turned against itself, and instead of focusing on production, society is more and more preoccupied with controlling the dangers (many of them involving environmental hazards) it produces. Another product of this modernization process is that risks have now become global dangers. Though the materialized consequences of risks may be perceived locally, their origins have become deterritorialized and are not easily traceable.[17] Furthermore, contemporary risks, difficult to be understood because they often involve complicated physical and chemical formula, are neither perceptible nor tangible at all (Beck, *Risk Society* 28). They are often not locally perceptible or only at a later point in time. The latency of these risks then further amplifies their deterritorialization. This does not imply that risks are not there: "Their invisibility is no proof of their non-existence; instead, since their reality takes place in the realm of the invisible anyway, it gives their suspected mischief almost unlimited space" (Beck, *Risk Society* 73). In this space of uncertainty and speculation, risks are omnipresent and yet not localizable and thus proliferate even more, as society refrains from stopping what may endanger it. As Frederic Buell, building on Beck, elaborates,

> In early modernity, the problem was scarcity, and the production and distribution of wealth were the chief focus. In advanced or late modernity, "overabundance" and "overweight" are the problems, and the production and distribution of risks become the more salient activities. In risk society, overabundance of development leads to an overabundance of man-made risk. (194)

The undelimitability and imperceptibility of these newly produced risks of modernity—and their inherently speculative nature, as Buell points out—make it difficult to raise awareness regarding their seriousness. Moreover, in a society in which risks are omnipresent, apocalypticism may no longer serve to convey a sense of urgency. Instead, "The primeval has disappeared, and the future will bring no novel apocalypse, but an immensely heightened version of old woes getting worse and still worse" (F. Buell, "A History" 29). Heise has pointedly summarized Buell's argument: while apocalypse offers the possibility of an alternative "risk-free" future, for a risk society that dwells in crisis, future environmental hazards can only be mitigated but never completely prevented (*Sense of Planet* 142). Moreover, apocalyptic imageries generally based on presenting clear antagonists, may no longer be fitting to grasp the diffuse relation between victim and culprit in contemporary risk scenarios. As explained in the foregoing discussion, Buell sees the more "realist" environmental declensionist narrative as well as the risk narrative as suitable frameworks to communicate the environmental crisis of modernity. Though agreeing that apocalyptic gloom and doom rhetoric may be problematic, I would question Buell's argument that there is only one narrative template that best communicates today's environmental problems. On the contrary, the complexity of today's risks requires writers, filmmakers, and activists alike to come up with new and innovative ways of portraying environmental risks.

Climate change epitomizes the challenges of communication that contemporary environmental crises pose. It is a problem of such complexity and so intricately intertwined with the current, globalized functioning of the world that it becomes impossible to point to its singular source of origin or to one particular weather extreme as its outcome. Because of their systemic character, climate risks seem overwhelming and all-encompassing, yet, at the same time, not graspable. The latency of climate risks, both temporally and spatially, further complicates our immediate perception of climate change. How then can the geographically distant impacts of climate change be brought closer to home? And how can the effects of climate change on future generations figure in today's thinking and decision-making? These questions, of course, also point to issues of climate justice. However, the factor of uncertainty inherent in future scenarios about climate risks has so far often served as a recipe for non-action instead of thinking about a future that can be responsibly shaped. Yet, this task requires people to understand themselves as an intricate part of the environment and acknowledge that while they may not be able to see the causes of climate change, they are originating from humankind's large-scale intervention in nature. Moreover, climate change serves as an example for what Rob Nixon has called "slow violence." He uses the term to describe a variety of incremental, long-term environmental problems such as climate change and acidifying oceans, the toxic and radioactive aftermaths of wars, and deforestation (Nixon 2); these are all processes that go beyond the "spectacular" of apocalypse, and work as a much "quieter" but no less violent crises.

While many environmental activists, nonfiction authors, and film directors still rely on apocalyptic rhetoric when addressing issues of climate change, there has been a growing repertoire of climate change literature that provides a more nuanced approach to this multifaceted environmental problem.[18] As literary scholars Adeline Johns-Putra and Adam Trexler suggest, "The sheer complexity of such a phenomenon … provokes and inspires authors to invent and innovate, in order that they can adequately imagine, depict, and convey it" (Trexler and Johns-Putra 185). Indeed, most of the climate change fictions discussed in this book go beyond apocalyptic rhetoric. While these works depict a climatically altered world, this world itself is not ending. Instead, writers implicitly seem to concur with Garrard's argument that: "Only if we imagine that the planet *has* a future, after all, are we likely to take responsibility for it" (*Ecocriticsm* 106). Climate change fiction may thus figure as an innovative response to the changed nature of environmental problems today and provide insights into alternative narrative frameworks to the apocalyptic environmental crisis discourse that has predominated until recently.

Ecocriticism, Climate Change, and the End of Nature

The discipline that foremost engages with the study of the "relationship between literature and the physical environment" (Glotfelty and Fromm xix), and thus the discipline that would be interested in the narrative frameworks of climate change, is ecocriticism. Before situating the topic of climate change within ecocriticism, it is important to briefly elaborate on the particularities of this "avowedly political mode of analysis" (Garrard, *Ecocriticism* 3). Just as climate change prompts new ways of envisioning today's environmental problems, it also challenges ecocritics "to something of a readjustment of [their] approaches and methods" (Trexler and Johns-Putra 185).

Only recently have ecocriticism and green cultural studies made their transition from marginal subjects to recognized areas of research in the humanities. As scholars such as Noel Sturgeon point out, issues of environmentalism were starting to have an impact on literary and cultural studies in the late 1980s, but were only becoming visible as an institutional presence at the turn of the millennium.[19] Unlike other social movements of the '60s and '70s, which had heavily influenced literary criticism, environmentalism only started to resonate with literary studies long after environmentalism had already reshaped the country's political climate (Heise, "Teaching Ecocritical Theory"). Heise sees the reason for this delay in environmentalism's deviation from the prevalent conceptual framework of literary criticism at the time.[20] When the study of literature and the environment was finally acknowledged in literary criticism, environmentalism was already a very diverse and contested field, and ecocriticism came to reflect this diversity.[21] This diversity of approaches to the topic has also made it very difficult to subsume the corpus of works under one particular definition, and as Lawrence Buell points out, there has been a lack of "the kind of paradigm-defining statement [in ecocriticism] that, for example, Edward Said's *Orientalism* (1978) supplied for colonial discourse studies" (*The Future* 11).

Furthermore, early ecocritics were continuously grappling with the problem of positioning themselves between "nature-endorsing" or "nature-skeptic" (Soper 4), between materialist or constructivist, between deep ecologists or postmodernists—and trying to come to terms with the controversy generated by the "science wars." As Heise summarizes:

> Ecocriticism, with its triple allegiance to the scientific study of nature, the scholarly analysis of cultural representations, and the political struggle for more sustainable ways of inhabiting the natural world, was born in the shadow of this controversy. Even though the grounds of the debate have shifted since then, the underlying issues of realism and representation that informed the science wars continue to pose challenges for ecocritical theory. ("The Hitchhikers Guide" 506)

Though ecocriticism is still lacking consensus on the concepts and ideas at its core, there have been several publications that aim to define the field and integrate its diverse formations into one body of knowledge. Great contributions include the *Ecocriticism Reader* (1996) by Cheryll Glotfelty and Harold Fromm, Greg Garrard's *Ecocriticism* (2004), and most recently Timothy Clark's *The Cambridge Introduction to Literature and the Environment* (2011). These publications provide an overview of the key texts in ecocriticism and illustrate the variety of environmental issues at stake.[22] They also call for cultural and literary studies scholars to engage more closely with the environmental sciences, for more interdisciplinarity, and for an engagement with an analytic canon that goes beyond literature. The influential trilogy in environmental cultural criticism by Lawrence Buell—*The Environmental Imagination*, *Writing for an Endangered World*, and *The Future of Environmental Criticism*—is of further importance, as it gives an insightful overview of the terminology and the development of the discourse of the discipline over the last decades. In particular, Buell elaborates on the idea that ecocriticism has helped to open up new textual archives formerly neglected in cultural studies, especially since it moved away from the sole analysis of nature writing and opened up to issues of environmental justice (L. Buell, *The Future* 22), thereby asking questions about the connection between environmental problems and social and cultural issues concerning race, class, and gender.[23] Gersdorf and Mayer's definition of ecocriticism also acknowledges these more recent tendencies. Drawing on Bateson's ideas in *Steps to an Ecology of the Mind*, they define an ecocritical project as "one that investigates aesthetic *representations*, discursive *performances*, and cultural *functions of nature* in historically, racially, and socially diverse communities and societies" (16).

Particularly for American Studies scholars, ecocriticism in its early stages was bound to deal with the seminal American nonfiction texts by transcendentalists Ralph Waldo Emerson and Henry David Thoreau and later works by John Muir, Aldo Leopold, and Edward Abbey. These texts derive from a tradition of thought that focused on a sense of place and an ethics of proximity based on local knowledge as well as an appraisal of the "wild" as the space of some pristine, untouched nature.[24] While the importance of these early texts cannot be neglected, they need to be read critically—acknowledging their conceptual limitations and shortcomings.[25] Too often they have been postulated as the ultimate texts of reference in ecocriticism, but the appeal to one authentic and sublime nature, so prevalent in this pastoral writing, may give way to dangerous political projects that emphasize the "superiority" of *one* particular kind of nature (Soper 32–33).[26] So the shift in "second-wave ecocriticism" (Buell, *The Future* 22) toward a broader and more critical understanding of environment was an important and necessary development in ecocriticism. As minority communities were exposed to increasing environmental risks, the formation of environmental justice criticism helped to call attention to the "toxic discourses" (Buell "Toxic Discourse") that emerged as a response to these contaminated (often urban) sites. Scholar and environmental activist T.V. Reed poignantly clarified that "aesthetic appreciation of nature has not only been a class-coded activity, but the insulation of the middle and upper classes from the most brutal effects of industrialization has played a crucial role in environmental devastation" (quoted in Heise, "Hitchhiker's Guide" 508). This "second-wave ecocriticism" thus also served to critically re-evaluate the discipline itself. Today, ecocriticism has opened itself up to a variety of texts, and also to different forms of critique, such as postcolonial and poststructuralist theory, as well as to a variety of environmental problems such as waste, loss of biodiversity, toxicity, and to works from different national contexts. Moreover, Ursula Heise's groundbreaking publication *Sense of Place and Sense of Planet* most notably challenged ecocriticism's overemphasis on rootedness in place and called attention to global environmental visions.

Yet, until the late 2000s ecocriticism neglected one of the today's most pressing environmental issues, climate change. In the introduction to his book on literature and the environment, Timothy Clark offers several explanations for the dearth of literary criticism concerning climate change. He explains, "As a global catastrophe arising from innumerable mostly innocent individual actions, the issue does not present an easily identifiable or clear-cut political antagonist. [Moreover i]ts causes are diffuse, partly unpredictable and separated from their effects by huge gaps in space and time" (11). Adding to Clark's argument, I would point out that ecocriticism's longstanding lack of engagement with climate change is also a result of a "crisis of the imagination," which has affected authors and artists alike in their attempts to represent anthropogenic climate change. Even most of the popular nonfiction books that deal explicitly with anthropogenic climate change, apart from Bill McKibben's early publication *The End of Nature* (1989), were only published after the turn of the millennium. These include Tim Flannery's *The Weather Makers* (2005), James Hansen's *Storms of my Grandchildren* (2009), and parts of Naomi Oreskes's *Merchants of Doubt* (2010). Interestingly, many of these nonfiction works on global warming rely heavily on apocalyptic rhetoric to give their words more urgency. An analysis of literary techniques employed in these nonfiction texts would certainly prove to be insightful. Yet, this book focuses on another category of texts, climate change fiction, which is only slowly starting to receive attention by literary scholars. In 2005, Robert Macfarlane, writer and literary critic, asked the "burning question" in the online culture section of *The Guardian*: "Where are the novels, the plays, the poems, the songs, the libretti, of this massive contemporary anxiety [of climate change]?" Throughout the article, he further argues that the mass of "invisible literature"—a term coined by J.G. Ballard that refers to "the data buried in company reports, specialist journals, technical manuals, newsletters, market research reports, internal memoranda"—is not enough to communicate the most pressing issue of our times, and thus needs to be supplemented by cultural imaginaries. This imaginary repertoire is of tremendous importance, Macfarlane stresses, in order to communicate the magnitude of climate change, to evoke emotions about it, and to ignite a debate about actions against its devastating effects.

Certainly, there are several literary works and films that engage with phenomena related to global warming, and large-scale climatic changes figure prominently in a variety of books. In science fiction, for example, there are references to extreme droughts in works such as Frank Herbert's *Dune*, to floods in works such as J.G. Ballard's *A Drowned World*, and to rising sea levels as in works such as Ursula Le Guin's *Always Coming Home*. In addition, Cormac McCarthy's *The Road*, Jennifer Egan's *A Visit from the Goon Squad*, and Margaret Atwood's *Oryx and Crake* and *The Year of the Flood* draw on climatic changes.[27] However, taking a more narrow definition of climate change fiction as a starting point, I am interested in works that explicitly engage with anthropogenic climate change. In these books, meteorological phenomena do not just provide the background setting against which the story unfolds; climate change significantly alters and is a prevalent issue for characters, plot, *and* setting. The climatically changed planet itself thus becomes an indispensable part of the narration. Bill McKibben, in his introduction to the short story collection *I'm With the Bears*, stresses this important characteristic of works of climate change fiction: "Instead of being consumed with the relationship between people, they increasingly take on the relationship between people and everything else" (McKibben, "Introduction" 3–4). While portraying humans' intricate relationship with the planet and thereby teasing out the anthropogenic dimension of global warming, climate change fiction inevitably also deals with ethical questions about humankind's responsibility in this unprecedented crisis. Finally, in depicting climate change, writers also have to engage with a variety of representational challenges such as the phenomenon's global scale and long, slow process of unfolding.

Departing from this more narrow definition, a growing corpus of climate change fiction has been published since the turn of the millennium.[28] In addition to well-known filmic examples such as the blockbuster *The Day After Tomorrow* (2004), two climate change books received significant attention from a wider audience: Michael Crichton's controversial but bestselling thriller *State of Fear* (2004), which deals with the uncovering of climate change as a scam made up by environmentalists, and the more recently published and media-hyped novel *Solar* (2010) by British novelist Ian McEwan, which draws an allegory between the rise and fall of its protagonist and humanity's inaction on climate change. Still the number of works is far from abundant, and since this publication focuses on American climate change fiction, the selection of texts for in-depth analysis was narrowed down even further.[29] However, looking beyond a scholarly bias in literary studies, generic definitions and presupposed categorizations of "high" and "low" literature, a small canon of works engaged with this global phenomenon of climate change can be identified. It includes thrillers like Matthew Glass's *Ultimatum* (2009); science fiction novels like Norman Spinrad's *Greenhouse Summer* (2000), Dale Pendell's *The Great Bay* (2010), and Paolo Bacigalupi's *The Windup Girl* (2010); short story collections such as *I'm With the Bears* (2011, edited Mark Martin) and *Welcome to the Greenhouse* (2011, edited by Gordon van Gelder). Finally, and most importantly, it also includes the books chosen for this analysis: Barbara Kingsolver's *Flight Behavior* (2012), Steven Amsterdam's *Things We Didn't See Coming* (2011), T.C. Boyle's *A Friend of the Earth* (2004), Jean McNeil's *The Ice Lovers* (2009), Kim Stanley Robinson's *Science in the Capital* trilogy (2005–2007), Dana Stein's *Fire in the Wind* (2010), and Octavia Butler's *Parable of the Sower* (1993).[30]

As the year 2015 marked another record high in CO_2 emissions and global warming is further on the rise, ecocritics but also scholars from other disciplines have started to respond to this particular environmental crisis in growing numbers. As early as 2008, Ursula Heise's chapter on the global dimension of climate change literature and film in her book *Sense of Place and Sense of Planet* (2008) has made an important theoretical contribution to the field. Some of the other early essays by ecocritics include Mike Ziser and Julie Sze's article on "Climate Change, Environmental Aesthetics, and Global Environmental Justice Studies" (2009), which identifies several conservationist and preservationist trends in cultural texts on climate change; the extensive overview of climate change literature in "Climate Change in Literature and Literary Criticism" by Adam Trexler and Adeline Johns-Putra (2011); as well as Johns-Putra's more specific essay on "Ecocriticism, Genre and Climate Change" (2010), in which she calls for a reconsideration of genre theory for ecocritical analysis and applies this to her own reading of Kim Stanley Robinson's capital trilogy. While it comes from a different disciplinary background, geographer Kathryn Yusoff's work on the political aesthetics of climate change also offers great insights on how the arts function as sites of political possibility and how the aesthetic experience enables us to understand environmental transformations (Gabrys and Yusoff; Yusoff). Especially in last couple of years, ecocritics and scholars from other disciplines have diversified the approach to literary and cultural productions concerned with climate change.[31] While Greg Garrard devotes a chapter of his guidebook *Teaching Ecocriticism and Green Cultural Studies* (2012) to the pedagogical aspects of climate change, Timothy Clark approaches the issue of climate change and ecocriticism from a deconstructivist perspective and finally concludes that climate change ecocriticism so far plays into the "simulative politics" it seeks to criticize (T. Clark, "Some Climate Change Ironies: Deconstruction, Environmental Politics and the Closure of Ecocriticism" 147).[32] Again others such as environmental justice scholars Janet Fiskio or Julie Sze have exposed the universalizing tendencies of climate change discourses, arguing that global warming is likely to aggravate social inequalities based on race, gender, and class differences (Sze, "The Question of"; Fiskio "Apocalypse and Ecotopia").[33] Finally, in 2015 the first monograph on Anglophone climate change fiction was published. In *Anthropocene Fictions* Adam Trexler traces a longer history of Anglophone climate change novels, analyzing their literary form as well as their relation to science, politics and economics.[34] Different to my own approach, Trexler departs from a wider understanding of climate change fiction and includes novels that do not necessarily deal with anthropogenic climate change.

Most ecocritics dealing with cultural representations of climate change have had to move beyond a generic bias prevalent in literary criticism. Many of the texts that engage most overtly with climate change come from so-called "genre fiction"—that is, speculative fiction but also thrillers. As Johns-Putra explains, "The dramatic and emotional contours of climate change have to do with the future … This leads, usually but not inexorably, to genres that have to do with future worlds" (749). She thus alludes to a genre that so far has often been neglected by literary scholars and particularly ecocritics, but which shows an explicit preoccupation with the future: science fiction (sf).[35] While some of the climate change fiction discussed here would traditionally be marketed as science fiction, such as Kim Stanley Robinson's trilogy, there are several other authors who engage with climatically changed futures and who are not necessarily known as science fiction writers. Yet, mapping out a climatically changed future, writers like T.C. Boyle, nevertheless, draw on generic strategies of sf and engage with the estranging aspect of the *novum*, a key characteristic of science fiction. Defined as "a narratological megatrope, a figural device that so 'dominates' (Suvin's term) its fiction, that every significant aspect of the narrative's meaning can be derived from it" (Csicsery-Ronay 49), the climatically changed world in climate change fiction can be read as such a *novum* and thus would allow for a classification of these texts as science fiction or more generally speculative fiction. In their discussion of genre fiction and theory, Trexler and Johns-Putra, however, point out that the labeling of a book "depends not simply on the characteristics of that text, but on the identification of its author with the genre and the willingness of readers to read that text ... Science fiction, in particular, functions as a community of authors and readers, built on and defined by a sense of shared identity" (187). This may serve as an explanation as to why many authors who employ *nova* in their texts are not identified as sf writers by readers or critics (ibid.).

Yet, as climate change is blurring the boundaries between science fiction and other literary genres, the generic labeling implemented by market structures seems rather restrictive. Many climate change fictions published so far illustrate that hybridized forms of the novel (i.e. a blending of genre fiction and novel, quasi-documentaries, start and stop narratives) are employed and seemingly most suitable to deal with the representational challenges that climate change poses. Also Adam Trexler concludes, "novels about the Anthropocene cannot be easily placed into discreet generic pigeonholes" (*Anthropocene* 14). Recently introduced terms such as "cli-fi" by Dan Bloom, which appeared in several online articles, in climate change fiction sections of bookshops and is the title of an active Facebook community, thus may bear a reductive connotation. Carolin Kormann convincingly argues that because of its linguistic proximity to "sci-fi," the term "cli-fi" "makes the genre sound marginal, when, in fact, climate change is moving to the center of human experience" ("Scenes from a Melting Planet").[36] While, indeed, a catchy name serves as a powerful tool to raise awareness for the issue, "cli-fi" may possibly be a misleading choice of name. Just as many climate change fictions could be characterized as speculative fiction since they invoke *nova* but are not perceived as such by their readership, others are marketed as science fiction but are, with their exceptional detail, committed to realism (ibid.), and closely bear on scientific data and real-world settings. Moreover, if indeed there is a new genre of climate change fiction emerging, it can so far be characterized by its common thematic focus on anthropogenic climate change rather than by its commonalities in narrative style. On the other hand, this hybridity may in particular be the genre's key characteristic.

It is certainly true that as an ecocritic, one is especially interested in the construction and meaning-making process of particular texts, and Johns-Putra correctly emphasizes that for this reason it is also important to consider the context or the environment in which a particular generic text is produced and to which it corresponds. She summarizes that it is ecocritics' task "to understand how a given text might refer to the environment or how any given discourse has operated upon nature" and therefore "our business is to understand the way in which cultural constructivism works" (746).[37]

Taking Johns-Putra's argument as point of departure, the climate change fictions discussed in this book, interestingly, move away from end-time visions. While some of the novels present wide-ranging disasters, and they predominantly draw on dystopian imagery, they are not necessarily apocalyptic and engage with worlds at risk rather than worlds destined to end.[38] As already discussed earlier, climate change fictions depict a future in which environmental crisis is omnipresent. The novels thus outline what would be commonly defined as dystopias—places and societies that are generally considered worse than current ones (Moylan and Baccolini 1). However, unlike traditional dystopias, which tend to end with the worsening of the depicted doom scenario, the open-endedness and generic ambiguity of climate change fiction impedes closure. It thereby underscores climate change's aporetic nature. Moreover, while writers engaging with climate change—the epitome of modern risk—depict how climate risks materialize into catastrophe, how this affects future societies, and how the characters in their books deal with the risk of always newly arising disasters, they also contemplate questions of "how a society with other values (less wasteful or competitive or materialistic) might have averted calamity" (Mizruchi 199). They ultimately suggest that different futures may be possible if a different course of action is taken now. Several of the novels can thus be read as critical dystopias. As Moylan and Baccolini explain, "Critical dystopias allow both readers and protagonists to hope by resisting closure: the ambiguous, open endings of the novel maintain the utopian impulse *within* the work" (7). Critical dystopias thereby avoid the radical nature of the dystopia while remaining open to political opposition, and explore ways of transformation and hope in dark times (Moylan and Baccolini 235). Apocalyptic rhetoric is counterproductive to the critical dystopian element that is also part of risk narratives, which is based on "indeterminacy, uncertainty and the possibility of a variety of different outcomes" (Heise, *Sense of Planet* 142).

So rather than calling on nature's demise, as most prominently done in McKibben's *The End of Nature*, or lamenting the loss of nature due to human intermeddling, climate change fiction challenges ecocritics to rethink their "discipline" and be open to news ways and discourse that may help to evoke an environmental ethics of responsibility.[39] Nature has not come to an end. On the contrary, climate change is proving to us that nature is still here: It is a "quasi object," a nature-culture hybrid, with which humans are deeply and inseparably intertwined, which humans shape but which has its own agency.[40] Climate change fiction thus serves as a reminder that a dualistic nature/culture thinking is obsolete. It prompts ecocritics to move out of their comfort zone of commonly held perceptions of what nature and human's place in nature might mean and proposes an understanding of nature in which human and non-human are co-constitutive. Fictional texts thus have much to offer readers and critics alike in conveying a different understanding of nature in times of climate change. Finally, drawing on Garrard, climate change can serve as an example for what he terms a contemporary "postmodern megahazards" and as such it calls for narrative alternatives "to established ways of presenting and containing environmental crisis. These would have to cope with indeterminacy, long timescales, complex problems of agency and responsibility and the postmodern problem of the unseen, unquantifiable cyborg risk" (*Ecocriticism* 149). Climate change fictions engage with these challenges, thereby encouraging ecocritics for a reading beyond the dichotomies of modernity and questioning their own disciplinary confines.

Being confronted with such a pressing matter as global warming, cultural critics and social scientists have called for new and innovative approaches to this issue in order to grasp, reflect on, and discuss the drastic global changes that we will face (see Leggewie and Welzer). As this book aims to show, climate change fiction serves as a cultural means to participate in climate change discourse, building on a concern for the environment without necessarily relying on apocalypse or elegy—lamenting the loss of some pre-human, pristine nature. Instead, it provides innovative means to narrate humankind's intricate involvement in climate and nature while showing that there is no escape from the climate risks in which we are already immersed. The authors of climate change fiction thereby inevitably engage with the challenges of time, place, and human agency that climate change poses.

Reading climate change fiction not only for its narratological relevance but also for its contribution to and critical intervention in climate risk communication, the authors of these works could be considered as (artistic) whistleblowers in and of a society that increasingly dwells in crisis. Whereas once doomsayers "sought to reveal awful truths to ignorant people and urged immediate action to avoid disaster, now voices need to ask people to acknowledge what they already suspect and what their society, even when denying environmental crisis, is still preoccupied with" (F. Buell 202). Writers urge readers to take environmental crisis seriously, thereby going beyond the contemporary "negative desire for apocalypse" (Swyngedouw 219) for entertainment purposes by setting their works in a future, in which global warming can no longer be ignored but constitutes an important aspect of everyday life.[41] Finally, ecocritics who engage with climate change fiction can make a valuable contribution to discussions concerning climate change in trying to point out how these narratives not only mediate information about climate change, but also shape our very understanding of it.

Notes

1. Average global temperatures so far show a warming of 0.8 °C in the years from 1880 to 2012 (IPCC, *Climate Change 2013* 3). Many industrialized countries have agreed that a rise of global average temperature of more than 2 °C (4° Fahrenheit) is to be avoided, because the consequences are otherwise expected to be severe (Archer and Rahmstorf 224–226).
2. The IPCC is an intergovernmental scientific body which "reviews and assesses the most recent scientific, technical and socio-economic information produced worldwide relevant to the understanding of climate change" (IPCC, "Organization"). It is organized into three working groups responsible for: the science of climate change, the potential impacts, and ways to adapt to climate change and the potential to reduce GHG emissions. The reports that the IPCC has produced so far are widely considered to be the "authoritative statements of scientific knowledge about climate change" (Dessler and Parson 24).
3. Already as early as in the 1820s, French physicist J.B. Fourrier postulated that a "layer of air" regulated the earth's temperature. He tried to explore the different "greenhouse gases" that were responsible for trapping heat and thus contributed to the "greenhouse effect" (see also Archer and Rahmstorf 8).

4. Awareness of climate change also increased due to discussions about the hole in the ozone layer over the Antarctic and the need to take action against further depletion, which ultimately led to the issuing of the Montreal Protocol (1989). The recognition that humankind might be leading to the destruction of a fragile planet finally culminated in the designation of "Endangered Earth" as the "Planet of the Year" instead of the "Person of the Year" by *Time Magazine* (1988) (Dessler and Parson 22).
5. The "common but differentiated responsibilities" principle is based on the idea that all nations need to address the climate crisis and are responsible for mitigation measures, but it acknowledges different levels of economic development and calls for 'developed' countries to take greater responsibility.
6. The Kyoto Protocol, though issued in 1997, only came into effect in 2005, once it had been ratified by enough nations, making up at least 55 percent of GHG emissions by industrialized countries in 1990. Over the commitment period from 2008 to 2012, emissions were supposed to be reduced to 5.2 percent below 1990 levels if all nations had complied with their reduction targets (Dessler and Parson 26). Up to 2015, however, even though some nations with Kyoto targets have successfully reduced their individual CO_2 emissions, worldwide emissions have increased.
7. The Clean Power Plan involves the reduction of carbon dioxide emissions from power plants by 32 percent from 2005 levels by 2030 (see EPA).
8. Partisan differences are especially prominent in recent polls on climate change. Only 14 percent of Republicans responded that global warming is a "very serious" problem (compared to 50 percent of Democrats), and only 16 percent of Republicans even believed that there is solid evidence that the earth is warming due to human activities (compared to 53 percent of Democrats) (Pew Research Center, 2010 in Giddens *Politics of Climate Change* 106).
9. What came to be known as the "hockey stick" graph powerfully visualized the rise in average global temperatures over the past one thousand years and showed a sharp increase in temperature in the second half of the twentieth century. The graph was ultimately used to prove the correlation between rising temperatures and growing CO_2 emissions.

10. Glenn Scherer points out that "roughly 50 million right-wing fundamentalist Christians in the United States believe in some form of End-Time theology" ("Christian-right Views are Swaying Politicians") and that they have had significant influence on Republican politics in the US.
11. Carson's book, which sought to convince her readers of the dangers of pesticides, is generally considered as the founding text of the modern environmental movement. It cannot be neglected that its great popularity correlates with the particular social context at that time. Young Americans in particular were increasingly dissatisfied with what they perceived as the evils of contemporary American life: the repression of African Americans, the war in Vietnam, corporate exploitation, among many other issues (Flippen 276). So as human and natural environments were both put under pressure by urban and industrial developments, the reduction of environmental hazards was just one of the many changes that Americans demanded. Nevertheless, the impact of Carson's book is not to be underestimated: her publication strengthened the environmental movement, culminating in the first Earth Day on April 22, 1970.
12. Frederic Buell points to Carson's work as one of the founding publications of apocalyptic environmental discourse. He explains, "She translated—with a prophetic urgency and ardour still powerful today—nuclear holocaust into an apocalypse bought about by the 'mysterious blight' of the 'white granular powder' (in reality, chemicals like DDT) falling on fields, silencing life there" (F. F. Buell, "A Short History of Environmental Apocalypse" 14). In his article on toxic discourse, ecocritic Lawrence Buell further emphasizes that nuclear fear figures as an important impetus at the time of Carson's work, not only for the environmental imagination but also in popular culture, as his examples of Paul Theroux's *O-Zone* or Philip K. Dick's *Do Androids Dream of Electric Sheep?* show ("Toxic Discourse" 648).
13. Fredrick Buell provides an extensive and very insightful overview of "apocalypse" in environmental discourse in his book *From Apocalypse to Way of Life*. While he concludes that apocalyptic rhetoric has become part of everyday life and thus might no longer function as a fitting trope to communicate environmental crisis, Palmer and Killingsworth in their equally informative article "The Apocalyptic Narrative from Silent Spring to Global Warming" argue that apocalypse, nevertheless, still functions as a powerful rhetorical device.

While this section focuses on *environmental* apocalypse, it is, however, important to note that eschatological themes and language have generally played an important role in Western societies. Among the main features of the Judeo-Christian apocalyptic tradition that come to be modified in environmental discourse, Buell names: "sudden rupture with the past, presentation of a revelation, narration of a world-end and dramatization of a last judgment" ("A Short History" 15).

14. Since apocalypticism is inherently proleptic, as Garrard explains, it is always responding to crisis on the one hand and feeding into the construction of crisis on the other (*Ecocriticism* 86).
15. The emerging scholarship in the humanities reveals a shift in academia. Based on the assumption that environmental science and politics alone are not enough to address environmental problems, environmental humanities try to "to enrich environmental research with a more extensive conceptual vocabulary" (Rose et al. 2) and question the category of the human especially in relation with its nonhuman environment. Next to this scholarly shift, growing environmental concerns also reached a wider public audience through an increase in popular eco-documentaries about a wide range of environmental problems such as water scarcity in *Flow*, hazardous resource extraction in *Crude*, the exploitation in the agrarian sector in *Food, Inc.*, and climate change in *An Inconvenient Truth*, among many others. Moreover, while environmental crisis has continuously played an important role in science fiction, it also began to figure more prominently in novels, for example, by T.C. Boyle, Jonathan Franzen, Don DeLillo, Cormac McCarty, and Barbara Kingsolver.
16. See also Charles Perrow's analysis in his seminal work *Normal Accidents*, in which he explains how even minor system accidents can lead to major disasters due to the increasing complexity of the underlying technological system.
17. Using the term "deterritorialized," I am not referring to Gilles Deleuze and Felix Guattari's usage of the term so much as to a more literal meaning—, i.e., the detachment of an object or cultural artifact from its place of origin on earth. See also my discussion in the following chapter.
18. For a brief overview of apocalyptic rhetoric in the media, see also Killingsworth and Palmers' article. In the following section, I will also provide some nonfiction examples of apocalyptic climate change literature.

19. Ecocriticism, is usually identified as beginning in 1992 with the founding of ASLE—the Association for the Study of Literature and the Environment (Clark, Glotfelty). It is generally acknowledged that there were many ecocritical articles published before the 1980s, but they tended to be seen as individual writings and not as part of a wider mode of analysis.
20. In her introductory essay "The Hitchhiker's Guide to Ecocriticism," Heise elaborates that one of the reasons for late development of ecocriticim was the close engagement of literary critique with French (post)structuralism during the late '60s until the early '90s, which aimed at demonstrating the disjuncture between meaning/representation and some outer reality as well as at a denaturalization (505). She points out that "this perspective obviously did not encourage connections with a social movement aiming to reground human cultures in natural systems and whose primary pragmatic goal was to rescue a sense of the reality of environmental degradation from the obfuscations of political discourse" (ibid.).
21. In the introduction to his volume on ecocriticism, Timothy Clark refers to the deep schisms that divide environmentalism. He points to the differences between reform environmentalists, who consider nature mainly as a resource that nevertheless needs to be defended against over-exploitation; deep ecologists, whose biocentrism claims the intrinsic value of all natural life; and finally, social ecologists, who see environmental problems foremost in relation to more general social problems (1–2).
22. There are a several other pivotal texts, such as Karla Armbruster and Kathleen Wallace's *Beyond Nature Writing: Expanding the Boundaries of Ecocriticism* and Laurence Coupe's *The Green Studies Reader: From Romanticism to Ecocriticism,* which aim to provide an overview of recent developments in ecocriticism and green cultural studies. Kate Soper's *What is Nature?* (1995) has also become a seminal text for ecocritics, as it retraces the philosophical approaches to nature and culture over time.
23. Many of these "new" ecocritical texts have been put together in *The Environmental Justice Reader,* edited by Adamson, Evans, and Stein.
24. The conceptualization of nature as spaces of sublime otherness and as the "ultimate landscape of authenticity" (Cronon 80) has been proven problematic, most notably, in William Cronon's historical account on the concept of wilderness. As he argues, environmentalists' claim for

the preservation of wild spaces, especially in the United States, often takes on an ahistorical perspective and neglects that theses landscapes were not an "original" and "pure" product of nature were instead the product of cultural processes, which often included a forceful expulsion of natives from these "uninhabited spaces." Furthermore, alluding to the eco-tourism that these places of wilderness prompted, Garrard summarizes the "trouble with wilderness": "At best, the wilderness experience and its deep ecological philosophy risks the identification with privileged leisure pursuits that sell authenticity while mystifying the industrialised consumerism that makes them possible" (71).

25. Garrard argues that the pastoral idea of wilderness will continue to be of importance, especially in the US context, because it "continues to supply the underlying narrative structure in which the protagonist leaves civilization for an encounter with non-human nature, then returns having experienced epiphany and renewal" (49).
26. Moreover, eco-feminists have argued that the tendency in ecocriticism to uphold nature/culture and nature/human divisions implies a further *logic of domination*—the prior always being inferior to the latter, which legitimized its exploitation and discrimination—that is that "ultimately, the human subject was defined as male and was given privilege over a feminized nature and over women and other social and racial/ethnic groups which were defined as belonging to the realm of nature" (Mayer 114).
27. For a thorough and extensive overview of climate change (both anthropogenic and non-anthropogenic) and climate change related phenomena in Anglophone literature, see Trexler and Johns-Putra's insightful article "Climate Change in Literature and Literary Fiction" as well as Trexler's recent publication *Anthropocene Fictions.*
28. Yet, it should be noted that Richard Turner's *Drowning Towers* was published in 1987, and Arthur Herzog's *Heat* was published even earlier, in the 1970s. These works are often cited as the first climate change novels (see Trexler and Johns-Putra).
29. There is a notable number of British writers—such as Jeanette Winterson, Marcel Theroux, Sarah Hall, and Helen Simpson—who have dealt explicitly with climate change in their writings in the last couple of years (see also Trexler and Johns-Putra 188). In addition, Franny Armstrong's crowd-funded docu-drama *The Age of Stupid* (2009) deserves mentioning here as a very innovative filmic depiction

of climate change, with a particular focus on its global dimension. Interestingly, climate change has found great resonance in children's and young adult fiction, such as the *Carbon Diaries* (2009) by British author Saci Lloyd or *2084—Nora's Welt* (2013) by the well-known Norwegian author Jostein Gaarder. Writers like Ilija Trojanow and Dirk Fleck have contributed to a growing canon of German climate fiction (see also Goodbody; Mehnert).

30. Even though I chose to focus on books, the last chapter includes an analysis of Paolo Bacigalupi's short story "The Tamarisk Hunter" from the collection *I'm With the Bears* as well as of Ben Zeitlin's recent movie *Beasts of the Southern Wild*, because these two climate fictions serve as particularly good examples for my discussion of climate justice. Finally, it has to be noted that while I am mostly discussing US fiction, I have included the Canadian author Jean McNeil and her book *The Ice Lovers* in my analysis. Her novel not only fits the criteria for selection, but also serves as an outstanding and innovative example for the telling of time in climate fiction.
31. I am purposefully naming here only publications that deal with literary representations of climate change. However, there are, of course, several ecocritics who have dealt with filmic representations of climate change such as Alexa Weik von Mossner in her articles "Facing The Day After Tomorrow: Filmed Disaster, Emotional Engagement, and Climate Risk Perception" and "Troubling Spaces: Ecological Risk, Narrative Framing, and Emotional Engagement in *The Age of Stupid*" or with global warming in more general terms, such as Scott Slovic in *Going Away to Think*.
32. In addition, there is an entire edited collection devoted to the topic of *Teaching Climate Change in Literary and Cultural Studies*, edited by Shane Hall, Stephen Siperstein and Stephanie LeMenager, which will be published in 2016.
33. The list given above is by no means exhaustive. There are many other publications of importance to the field, for example, Dipesh Chakrabarty's articles on climate change ("Postcolonial Studies and the Challenge of Climate Change"; "The Climate of History: Four Theses") or Rob Nixon's *Slow Violence* as well as numerous others that I draw on in the following chapters.
34. Others such as Danish scholar Gregers Andersen's dissertation deals with global warming in fiction and philosophy, and Janet Fiskio from the United States is working on a monograph examining climate change in art, performance, and protest.

35. Even though science fiction has often dealt with issues of environmentalism and "no genre potentially matches up with a planetary level of thinking 'environment' better" (Buell, *The Future* 57), it has tended to be neglected by ecocritics (Buell, *The Future* 56 ff). However, with an increased interest in popular culture and due to sf's innovative narrative structures—a blending of magical realism and cyberpunk—this has started to change (ibid.). Furthermore, since science fiction has become a part of mainstream images of contemporary popular culture, it has become difficult to be defined as a distinctive genre at all (U. K. Heise, "Introduction: The Invention of Eco-Futures" 2; Bould and Vint 182).
36. Korman, Bloom and several other journalists use cli-fi to describe an entire 'genre of climate change'. However, without going into further discussion of the characteristics of this particular genre, the employment of the term in this context may be oversimplified and prejudged.
37. Though I agree with Johns-Putra that the analysis of the mechanisms of cultural constructivism is at the methodological heart of our discipline, I question her quick dismissal of the material value of environmental problems. If ecocriticism is to continue as the highly interdisciplinary field it is today, ecocritics cannot emphasize solely constructivist aspects, but need to acknowledge material aspects as well discursive formations of the environment. I thus concur with Garrard's statement, which acknowledges that "'nature' is always in some ways culturally constructed ... and nature really exists, both the object and, albeit distantly, the origin of our discourse" (Garrard 10).
38. Heise has pointed to the importance of risk theory for the study of literature—not only for literature's thematic engagement with risks, but also for its influence on literary form. Furthermore, she argues, by studying how risk scenarios are made intelligible in fiction, ecocritics can contribute and broaden the discussion of environmental discourse (U. K. Heise, "Risk and Narrative" 748). Climate change fiction is inevitably a form of risk communication and as such makes for an interesting case study for the kinds of genres and frameworks employed to depict climatic risks (see also in particular my discussions of risk and literature in Chaps. "Scaling Climate Change: The Transformation of Place in Climate Change Fiction" and "Manufactured Uncertainty: Climate Risks in an Age of "Heightened Security""). While arguing that apocalypticism may fall short in

depicting contemporary environmental crises such as climate change, I am aware that it continues to play an important role in environmental discourse. Also some of the climate change fiction discussed in this book draw on apocalyptic imagery. Agreeing then with Heise, apocalypse and risk scenarios do "not amount to any fundamental dichotomy" (*Sense of Planet* 142).

39. In her introduction to the *Symploke* issue on "Critical Climate," Johns-Putra draws attention to the emergence of a new "critical climate" prompted by climate change, which circulates around the key questions if our current theoretical paradigms are apt to analyze a crisis such as climate change and conversely if climate change "conjures up old habits of critique" (Johns-Putra, "A New Critical Climate" 2). As Johns-Putra describes drawing on Clark: While the theoretical tools may remain the same (deconstructivist etc.), the object of analysis is nevertheless unprecedented. The critique is thus new, because it serves as a critical reorientation of the field of research (ibid.).
40. As Timothy Morton argued in *Ecology without Nature*, claiming that there is an authentic "original" nature is dangerous. Too often this conception of "pure" nature has been abused to legitimize nationalistic claims (15). Instead, Morton promotes an understanding of nature as something organic and vital, always within us, as something liquid and changing, constantly producing new meanings.
41. Philosopher Slavoj Žižek has pointedly summarized the current global predicament: "We all know about the impending catastrophe—ecological, social—, but we somehow cannot take it seriously" (Žižek). His example of an Austrian soldier who describes their situation during WWI as "catastrophic but not serious" serves to illustrate this condition. Žižek, ultimately, sees the problem as intricately intertwined with capitalism and calls for radical political and economic changes.

Scaling Climate Change: The Transformation of Place in Climate Change Fiction

Climate change is the epitome of a deterritorialized environmental crisis. With its diffuse relation between cause and effect, its latency and long timescale, as well as its global dimension, climate change eludes our perception even while it is omnipresent in media accounts and scientific reports. Whereas the effects of other environmental problems are often visible locally in the form of mine tailings, land erosion, or loss of biodiversity, the impact of climate change is cumulative and cannot be immediately felt. Moreover, so far it has been in the uninhabited areas of the world (the North and South Poles, high elevations, and coral reefs) where early signs of climate change have been detected. These geographically distant impacts "have to compete for attention with immediately felt physical needs, professional demands, economic necessities, or social obligations" (Moser 34)—often unsuccessfully. Contesting our focus on the local and going beyond national frames of meaning, an understanding of "global climate change" requires a new way of thinking—one that may involve the development of a global consciousness.

This chapter focuses on two climate fiction novels and the narrative challenge of grasping the spatial dimension of global climate change. By examining Steven Amsterdam's *Things We Didn't See Coming* and Barbara Kingsolver's *Flight Behavior*, it explores how a planetary transformation such as climate change impacts our understanding of locality and place and illustrates how these climate fictions ultimately present novel ways of

A. Mehnert, *Climate Change Fictions*,
DOI 10.1007/978-3-319-40337-3_3

envisioning climate change as the deterritorialized crisis it is. These novels acknowledge that climate change is a crisis caused by forces that transcend the local and thus can only be adequately understood from a global perspective. The literary works discussed here are innovative because they refrain from an advocacy of place and instead foster an awareness—not unlike Ursula Heise's call for an "environmental world citizenship" (*Sense of Planet* 10)—of the interconnectedness and mutual conditioning of human and non-human, natural and cultural places worldwide.

So far, our vision of climate change mainly derives from climatologists' abstract simulations. Moreover, in her analysis and critique of experts' abstract images in climate change discourses, Birgit Schneider explains that the problem of conveying the reality of climate change has mainly been one of scale:

> People observe daily weather changes, but they do not perceive climate—something which is, according to its modern definition, a statistically created abstract object of investigation with a long-term assessment period. Furthermore, people can only experience local weather, but not the global effects of climate change, which would require no less of them than to perceive the world as a whole.[1] (82)

Schneider points to the extreme difficulty of representing something as elusive as climate change and exposes the epistemological problem between expert knowledge and lay perception of climate change. Since the physical materiality of the issue is conceptualized through abstract statistical averages, global climate change becomes something that cannot be perceived or experienced per se. As Paul Edwards explains, "Only by coupling statistical analyses to climate modeling exercises have scientists been able to isolate and display the 'fingerprint' of global warming in changing weather patterns around the world" (33). Consequently, it was through computer modeling and graphic visualizations that climate change was first understood. For a large percentage of the population, climate risks exist mainly in the form of dangerous-looking graphs of future climate projections, but "the most commonly cited figure in climate change debates—change in the average global temperature—has no correlate in anyone's actual living conditions" (Edwards 33). As individual lived experience cannot serve as the basis for policy judgments that affect society, reliance on scientifically mediated climate risk scenarios becomes central in the decision-making process. As Ulrich Beck has argued, in modern societies "risk awareness is not based on 'second-hand experience,' but on 'second-hand non-experience'" as presented in various scenarios since "ultimately *no one* can know of risks if knowing means having consciously experienced them" (*Risk Society* 96).

However, the high level of abstraction involved in envisioning and understanding these scenarios often prevents us from seeing ourselves as actors, initiators, and victims of these risks. Unlike Edwards, who considers the scientific work on climate change to be the only basis for creating public awareness and understanding of global warming, I argue that there are other important and effective means, such as novels or films, to communicate the planetary transformations of climate change, especially in illustrating how a global and deterritoralized phenomenon such as climate change impacts our understanding of locality and place.

When dealing with complex issues, turning to cultural imaginaries like novels, and reading them alongside scientific writing, provides new insights because they reveal the intimate aspects of human struggles and bring formerly unacknowledged perspectives to light. Furthermore, as Jennifer Gabrys and Kathrin Yusoff emphasize, aesthetic practices function as "a way to materialize and articulate what would otherwise be unsayable and un-thinkable" (17). As discussed earlier, one prominent and "unthinkable" aspect of climate change revolves around its spatial dimension and people's perception and experience of this global phenomenon. Thinking about global climate change not only means considering what effects one's own lifestyle choices will have on distant places, but also imagining the ways in which global transformations will alter one's own local environment.

Ecocritics have yet to provide an in-depth exploration of literary works that attempt to grasp the spatial challenge of global climate change. Also Timothy Clark notes, "The relative absence in ecocriticism of its most serious issue seems more to do with the novelty and scope of the problem than with personal failing, a measure of how starkly climate change eludes inherited ways of thinking" (11). He therefore concurs with the argument Heise makes in her seminal work *Sense of Place and Sense of Planet*, by pointing out that so far many ecocritical discussions have relied heavily on an "ethics of proximity" and have thus focused on locality and place as the only position from which to articulate environmentalist critiques. Heise argues that environmentalists' ongoing emphasis on localism, their insistence on the need to reconnect with local places, has by and large ignored the ongoing discourse of "how experiences of place change under the influence of modernization and globalization processes" (51). Instead, Heise says that engaging with two key concepts from globalization studies, deterritorialization and cosmopolitanism, may provide a new and helpful point of departure for ecocritical analyses.[2]

While the concept of deterritorialization has been widely discussed by scholars from various disciplines (Deleuze and Guattari; Appadurai; Canclini; Tomlinson), they all define it as the detachment of socio-cultural practices from place, or more specifically, from geographically defined territories. Of great relevance for this chapter is the connection Heise draws between risk theory and Tomlinson's discussion of deterritorialization. She emphasizes that the global dimension of today's risks, such as climate change, underlines Tomlinson's argument that

> globalization promotes much more physical mobility than before, but the key to its cultural impact is in the transformation of localities themselves … This is in many ways a troubling phenomenon, involving the simultaneous penetration of local worlds by distant forces, and the dislodging of everyday meanings from their "anchors" in the local environment. (Tomlinson 29)

Accordingly, global risk scenarios, such as those involving climate change, may intensify processes of deterritorialization as people increasingly find themselves confronted with the transformation of localities and begin to question the value of certain places that have come to be endangered.

Indeed, the climate change novel discussed in the first part of the chapter, Barbara Kingsolver's *Flight Behavior*, illustrates the transformation of localities faced with global climate change. Kingsolver's book aligns well with Heise's concept of eco-cosmopolitanism, as it envisions "individuals and groups as part of planetary 'imagined communities' of both human and nonhuman kinds" (Heise 61).[3] *Flight Behavior* exemplifies that natural and cultural processes shape each other without regard for local or national borders, and thereby emphasizes that the local must always be viewed in relation with global processes. Only then can a global consciousness for climate change arise.

Steven Amsterdam's *Things We Didn't See Coming* also depicts the transformation of place by global climate change. However, his novel shows how places increasingly lose their socio-cultural meaning and are turned into *non-places* as climate change makes long-term habitation and attachment to place futile.[4] The analysis of Amsterdam's book furthermore explores how, consequently, in this future world affected by climate change, riskscapes emerge and roots become routes.[5] Hence, mobility and uprootedness are the only response to global climate change, which no longer allows for any risk-free spaces at all.

While *Things We Didn't See Coming* focuses on global uprootedness and *Flight Behavior* emphasizes global connectedness, both works tellingly demonstrate that a solely local perspective is no longer sufficient to grasp the global transformation of climate change. These two climate fiction novels make use of innovative narrative means in representing the spatial dimension of climate change and fostering a planetary understanding of the world.

Strange Flight Behavior: Climate Change, Butterflies, and Eco-Cosmopolitanism

The United Nations 2010 *Global Biodiversity Outlook* report not only concluded that the biodiversity targets defined in 2002 by the world's leading politicians were not met, but also that the rate of biodiversity loss since then has intensified, with climate change representing one of the principal drivers of the global decline in natural habitats. In addition, the Intergovernmental Panel on Climate Change (IPCC) has repeatedly pointed to the intensifying vulnerability of ecosystems due to rising global temperatures, very likely leading to the "disruption of species ecological interactions, and major changes in ecosystem structure and disturbance regimes" (IPCC in Archer and Rahmstorf 163). Even though these experts have indicated that effects of global warming on biodiversity are severe, to the majority of the world's population they have remained invisible so far. Many of the habitats in decline are found in isolated areas, and most humans, living in supposedly protected and artificially maintained city spheres, have difficulties seeing themselves as part of the larger ecological system. Furthermore, while the prominent media image of the drowning polar bear may evoke a sentimental reaction in most of us, the alarming demise of warm-water coral does not receive the same attention. There seemingly reigns then a double standard, where some forms of life are more likely to receive protection than others.[6]

Yet in times of global warming, we need to not only go beyond the division between which species are dear and close to us and which ones are not, but, more importantly, we need an understanding of our connection to and dependence on all forms of life. As Ban Ki-Moon has pointed out in response to the intensifying biodiversity loss,

> The consequences of this collective failure, if it is not quickly corrected, will be severe for us all. Biodiversity underpins the functioning of the ecosystems on which we depend for food and fresh water, health and recreation, and

> protection from natural disasters. Its loss also affects us culturally and spiritually. This may be more difficult to quantify, but is nonetheless integral to our well-being. (*Global Diversity Outlook* 3)

Destabilizations in the global ecological system will not only impact the availability of our basic resources but will also interfere with our cultural practices. Understanding the ways in which humans are linked to global ecology is paramount in dealing with climate change as well as the closely linked biodiversity crisis. It requires us, however, to imagine ourselves in a wider context of human and non-human belonging.[7] As John Muir, pioneer in wilderness protection planning and the founding father of the Sierra Club, explained, "When we try to pick out anything by itself, we find it hitched to everything else in the universe" (248). This statement acknowledges that the there is an interconnectedness of all forms of life and thus asserts that environmental phenomena are part of a larger, correlated, and global system.

As "new materialism" (Diana Coole and Samantha Frost)—which aims to acknowledge the agency, vitality, and self-constituency of all matter—is demarcating a paradigm shift in environmental humanities, discussions that revolve around non-binary subject-object relations and turn to new ways of representing the materiality of nature are gaining in importance. This trend also involves a decentering from the predominant human experience of and perspective on nature. In *When Species Meet*, Donna Haraway explains, "If we appreciate the foolishness of human exceptionalism, then we know that becoming is always becoming *with*—in a contact zone where the outcome, where who is in the world, is at stake" (244). Accordingly, the human is then just one agent in a larger non-hierarchical system of constant transformation and not the center that determines this "becoming."[8] For literature and literary criticism, this perspective implies an engagement with how non-human and human agency work together as a "choreography of becoming" (Coole and Frost 10) in the production of meaning. In the case of global warming, itself a "symphony" of co-emerging interactions, an awareness of the non-human is crucial to an understanding of the global scale of this phenomenon.

Turning to Barbara Kingsolver's recent novel *Flight Behavior* (2011), this chapter thus aims to explore the ways in which literary texts provide a realm to imagine ourselves as part of a *planetary* system of belonging.[9] The novel demonstrates that climate change requires thinking in multiple registers, especially if one is to make sense of locally occurring habitat changes.

The author juxtaposes the well-evolved migratory system of monarch butterflies to the strong sense of place among the human characters in the novel, and thereby innovatively questions our understanding of who figures as an agent of mobility in cultural and ecological globalism. The non-human perspective is here highlighted in the framing and narrating of global warming and species extinction. Moreover, *Flight Behavior* can be read in terms of Heise's eco-cosmopolitanism, illustrating that natural and cultural processes shape each other without regard for local or national borders. As the following close reading analysis will demonstrate, the human and the natural, the global and the local thus cannot be considered independently of one another but form part of planetary system of interdependencies.

The "Butterfly Effect"

The beginning of the novel *Flight Behavior* immediately juxtaposes the protagonist's ties to her home to the global transformations caused by climate change. Dellarobia Turnbow—who is suffocating in a life where even the "wide-screen version" includes nothing more than "the sheep in the field below, the Turnbow family land, the white frame house she had not slept outside for a single night in ten-plus years of marriage" (Kingsolver 2)—is determined to break free of this narrowly defined context and ready to commit adultery. However, an unusual spectacle awaits her on top of the mountain—the predetermined place for the sinful encounter. As the sunshine reaches the forest, the scenery begins to light up, seemingly setting the trees surrounding her on fire, leaving Dellarobia to watch the bright orange display of light in awe. This opening scene figures as a sublime moment in the narrative, and is worth being quoted at length here:

> She was pressed by the quiet elation of escape and knowing better and *seeing straight through to the back of herself*, in solitude. *She couldn't remember when she'd had such room for being.* This was not just another fake thing in her life's cheap chain of events, leading up to this day of sneaking around in someone's thrown-away boots. Here that ended. Unearthly beauty had appeared to her, a vision of glory to stop her in the road. For her alone these orange boughs lifted, these long shadows became a brightness rising. It looked like the inside of joy, if a person could see that. A valley of lights, an ethereal wind. It had to mean something. (15–16, emphasis added)

At first sight of this strange phenomenon, Dellarobia does not recognize that what she is witnessing is an entire overwintering population of monarch butterflies awakened by the warmth of the sun's rays. Though desperately trying to make sense of this encounter, she does not revert to religious explanations, "by no means was she important enough for God to conjure signs" (16), but rather sees in it a reflection upon her own life and being: a dangerous warning and a beautiful temptation at the same time. In its escape from "civilization," Dellarobia's epiphanic encounter with nature resonates with American pastoral narratives, but even more so with Romantic notions of the sublime, which indicated that nature's awe and beauty were essential for spiritual renewal. However, in *Flight Behavior* the close encounter with nature does not provide a moment of truth. The "nature" of the phenomenon on display is therefore neither revealed to the protagonist nor to the reader. Only when she visits the mountain a second time with a group of people does she realize that she had seen butterflies.

Yet, Dellarobia is not able to make sense out of the butterfly phenomenon up on the mountain because only a global perspective would enable her to do so. This is revealed at a later point of the story in her conversation with butterfly expert Ovid Byron when she states, "People can only see things they already recognize … They'll see it if they know it" (282). A locally occurring phenomenon, in this case the transformation of habitat, may be perceived locally, but it cannot be explained if not considered in relation to its larger global context. Dellarobia here refers to a major problem in mitigating climate change: the global dimension impedes the applicability of inference systems. Wondering why such a big monarch population has come to settle exactly on the Turnbow family property, Dellarobia finally consults the Internet—the one thing through which she is able to trace the invisible connections across the planet—and learns about the butterflies' wider migratory context. Monarchs follow a complicated transnational and multi-generational migratory system. Similar to birds, they are the only butterfly species that migrate north to Canada and south to Mexico in a multi-sequential system. In other words, while the first generation migrates north in the spring, their short lifespan prevents them from returning to their overwintering sites, and it is only the third or fourth generation will do so. As Ovid Byron explains to Dellarobia, "This system of local and universal genetics makes a kind of super-insect. The population can fluctuate fivefold in a year. It's an insurance policy against environmental surprises" (318). However, as the unusual overwintering site on the Turnbow property suggests, this seemingly secure system has now nevertheless been disrupted due to climate change, pushing the butterflies farther north and forcing them to leave roosting sites in Mexico earlier in the year (147).

The butterfly phenomenon up on the hill discloses how the human and the non-human together form part of one complex and interdependent system that climate change is threatening to disrupt. After learning about the monarch's disrupted flight behavior, the protagonist understands that what she has come to consider the "one spectacular thing in her life" turns out to be a "sickness of nature" (149). Making sense of the local phenomenon she experienced in a global context, she realizes that the "forest of flame that had lifted her despair" (229) was a transient moment of beauty, which revealed that the underlying "migratory pulse that had rocked in the arms of a continent for all time" (ibid.) is now threatened with extinction.

As Dellarobia starts to learn more about the butterflies during her work as a research assistant in Ovid Byron's lab, the multi-directionality of the ecological system begins to take shape for her. She understands that the butterfly phenomenon is "not just an orange passage across a continent as she'd imagined it before, not like marbles rolling from one end of a box to the other and back. This was a living flow, like a pulse through veins, with the cells bursting and renewing themselves as they went" (146). She realizes that what she is experiencing, though still not fully graspable, is not a static system but one that is generative and vibrant and one that affects her "with strong emotions" (ibid.). She suspects the greater meaning of this realization is to extend one's protection and feelings toward the non-human, as illustrated by her rhetorical question, "How was that even normal, to cry over insects?" (ibid.). Kingsolver's writing here illustrates the experience of the creative and self-directed power of nature, which allows for new sensibilities and an awareness of what Debora Bird Rose and Thom Van Dooren have described as "situated connectivities that bind us into multi-species communities" (87). Also Val Plumwood has argued that in current times of environmental crisis, it is crucial to be "open to experiences of nature as powerful, agentic and creative, making space in our culture for an animating sensibility and vocabulary" (46). Imagining ourselves in a wider planetary context requires this kind of openness.

While the butterfly phenomenon takes center stage in *Flight Behavior* and most prominently demonstrates the interconnected system between human and non-human actors, there are several other instances in the novel that emphasize bio-physical and socio-cultural interactions in the creation of habitats. As the novel continues, worsening climatic conditions put this system of interdependencies under increasing pressure. The Turnbow family (Dellarobia, her husband Cub, and her in-laws Bear and Hester) is dependent on the income from sheep farming and is now facing great financial debt due to their loss of hay crops because of increased rain levels. As a consequence, Bear Turnbow is planning to sign a contract with a logging company that wants to cut down the timber on the "butterfly" mountain behind Dellarobia's house. Dellarobia is not only worried about the implications for the butterflies, but also points to the dangers of a landslide that logging might provoke. Portraying numerous correlated factors, Kingsolver explores the complexity of climate change impacts. Animals, plants, the soil, and humans interact and together form the dynamics of the community, which shape and transform the place Dellarobia calls home. The novel thus powerfully demonstrates that the human cannot be considered by itself, but always stands in relation with its environment. Ovid Byron, the butterfly expert who sets up his research lab in the Turnbow family's backyard, explains that everything works in a series of interactions and "becomings": "'An animal is the sum of its behaviors,' he said finally. 'Its community dynamics. Not just the physical body'" (Kingsolver 317). A seemingly minor subplot revolves around the Turnbow family's neighbors, the Cooks, and subtly indicates that soil contamination may have caused their son to fall ill with cancer. As a consequence, they have turned to organic farming, which makes them subject to constant ridicule by their neighbors who believe cancer just like weather irregularities to be of divine intention. Their denial culminates in Bear Turnbow's trivial comment of storing DDT in his basement and feeling not the least worried about using it as well. When Dellarobia observes that Ovid Byron's life is actually "at the whim of a livid ecosystem" (394) this description could, in fact, be applied to all characters (human and non-human alike).

Hence, humans and non-humans in the novel are constantly set in relation to one another. This interrelation is also emphasized by the choice of character names in the novel: Cub, Dellarobia's husband, who does not think for himself but will always remain his parents' son; Bear, the bad-tempered and rather unyielding father-in-law; and Dellarobia's best friend Dovey, who is free as a bird and a source of hope. In addition to these rather anthropomorphic names, Dellarobia, "named after a wreath made out of nature junk" (105), represents the amalgamation of nature and culture, and symbolically foreshadows her recognition that everything is connected to everything else within a global ecological system. Dellarobia is then not only the name of an arts and crafts project, in which pinecones and acorns were glued to Styrofoam, but is also ironically homophone to a famous fifteenth-century Italian sculptor family name, Della Robbia, a name of international standing. Furthermore, the name Ovid tellingly alludes to the well-known Roman poet and his *Metamorphoses* (transformations), allegorically pointing not only to the transformative but also nature-culture bridging function of this character in the novel. The importance of names also extends to the monarch butterflies, known to the lay population as "King Billies." Tellingly, Ovid's wife Juliet, who is an anthropologist and not a biologist, is the only one who can explain to Dellarobia that the name came from early Protestant settlers who associated the orange colors of the butterfly with the royal colors of the English King William III. Once again, nature and culture, human and non-human creatures do not make sense if they are considered as detached from one another.

The employment of the term "butterfly effect" for the analysis is thus not in reference to chaos theory, but used more literally in order to explore the effect that the focus on butterflies in the story has—the emphasis on the relation between human and non-human and the importance of the global dimension—and which ultimately allows for an eco-cosmopolitan reading of the novel. Becoming aware of her own position in the larger context of butterfly migration, Dellarobia finally wonders on what basis communities and kinship systems are formed. Though she was born and grew up in this rural community, Dellarobia continuously feels "out of place" and this feeling of displacement is intensified when she is confronted with the displaced monarch butterflies that normally overwinter in Mexico. She starts questioning her attachment to the Turnbow property she lives on, "where she never belonged in the first place" (Kingsolver 358) and wonders, "what kind of ties were those, what did they bind?" (358). As the story unfolds, it becomes even more apparent that ties to one particular place are by no means "natural" and static, but are constantly shifting and defined anew. The novel then illustrates how questions of selfhood and identity are not only renegotiated in relation to the non-human, but also in relation to a large-scaled phenomenon such as global warming. Dellarobia summarizes, "Before this year she had hardly looked a butterfly in the face, and now they were star players in her own domestic drama. Which was officially no longer just domestic" (124).

Migratory Pulses

The butterflies' far-reaching migratory behavior stands in stark contrast to Dellarobia's own experience of the world. Dellarobia is fascinated "to think of these fragile creatures owning the span of a continent" (143), and realizes that these insects have crossed more borders that she ever has. She has never traveled on an airplane, never been out of state, and has rarely ever eaten foreign food. However, her close connection to the land and her dependency on it derives from financial necessity rather than from a pre-existing natural connection to the land, let alone from an environmental awareness. As a long tradition of environmental writing and criticism has tried to promote, attachment to locality and place is often essential to an environmental awareness. Working the land, according to prominent figures of New Agrarianism such as Wendell Berry, is a constituting aspect of one's identity and is crucial for the "cultivation of character, which in turn produces individuals capable of being responsible citizens" (Fiskio 302) within a clearly defined national context.[10] However, this overemphasis on "natural" place attachments not only neglects the ways in which "working the land" may also go along with exploiting the land and thus may have nothing to do with environmental consciousness. This perspective is also oblivious to how places increasingly become deterritorialized and are shaped by global agrarian market practices. As global practices (not only agrarian, but also media and consumer practices, for example) pervade locality, the importance of "place" for identity formation is further questioned.[11] Challenging former environmentalist discourses, Heise asks critics to acknowledge the consequences of modernity and "to shift the core of its cultural imagination from a sense of place to a less territorial and more systemic sense of planet" (Heise, *Sense of Planet* 56).

The intrusion of butterflies, global players in the novel *Flight Behavior*, into Dellarobia' s locally defined life further illustrates how a centering on a "sense of place" may be counterproductive for environmental awareness, especially in times of global climate change. In a world in which "everything is connected to everything else" (Commoner), an emphasis on place attachment and a sole focus on locality impedes valuable insights regarding the transformations taking place across cultures and national boundaries. As Dellarobia explains in a TV interview, "I was so focused on my own little life. Just one person. And here was something so much bigger." (Kingsolver 209). The protagonist admits that the small-scaled focus on her home had limited her. The butterfly event challenges this narrow vision and ultimately enables Dellarobia to see herself as part of a wider and interactive environment operating on multiple levels.

In addition to the appearance of the butterflies, several human encounters also add to Dellarobia's growing sense of planet and to her awareness of the global climate crisis. Indeed, it seems as if "the world is beating a path to [her] door" (110). The butterflies not only attract several out-of-state visitors, such as biologist Ovid Byron and his students or the activists of the grassroots organization 350 from California, but also confront Dellarobia for the first time in her life with people from other countries as well. Meeting the Mexican parents of her son's new friend, Josefina, she finds out that a landslide had destroyed people's habitat, as well as the monarchs' habitat, in Michoacán. Living off the butterfly tourism in the area, and having "*lost their world*, including the mountain under their feet and the butterflies of the air" (103, emphasis added), Josefina's family was forced to relocate only to find themselves reconnected with the monarchs that came to settle right on Dellarobia's property. What at first seems to be a locally occurring phenomenon quickly turns out to be an event of transnational ramifications. As the novel continues, further cross-cultural connections are drawn between the protagonist's life and the Mexican butterflies. Josefina tells Dellarobia that in Mexico, the monarch butterflies represent the souls of dead children. Learning about the meaning of the monarch in Mexican culture also helps Dellarobia to come to terms with the loss of her own first child. The novel here stresses not only the transformation of the local by the global due to physical mobility (of insects, people, etc.), but also due to the circulation of ideas and cultural meanings, which are disembedded from place and become adopted and reconfigured across the globe. Finally, this passage resonates Ban Ki Moon's statement about the cultural and spiritual significance of biodiversity presented in the beginning of this chapter, and serves as an example for environmental literacy and the ways in which an understanding of an ecological globalism or a global environmental consciousness may be communicated.

Kingsolver's description of the "Women Knitting the Earth" initiative further highlights how Dellarobia's place and life are no longer local, but are pervaded by global mechanisms. During one of her visits to the butterfly site up on the Turnbow mountain, the protagonist meets two women (mother and daughter) from England, who heard about the butterflies because of Dellarobia's internationally broadcasted TV interview, and decided to do a "sit-in against global warming" (299). As part of their grassroots campaign, they knit small butterflies from used orange sweaters that are being sent to them by people from around the world, and hang them up on the branches as a memorial for the butterfly population whose survival is endangered by global warming. The knitted butterflies tangibly and ideologically represent the community effort for action against climate change. They thus become symbolic tokens of a "disembedding mechanism" (Giddens), through which social practice—in this case action against climate change—becomes detached from its immediate and local contexts. Most importantly, place is here transgressed by "complex connectivities," by an "ever-densening network of interconnections and interdependencies" (Tomlinson 2). With the online campaign "Knitting the Earth," the two women bring together the threads of the global network, and make visible—through a *mise en abyme* of the actual phenomenon—how the local (Dellarobia's mountain) is pervaded by the global (the changed butterfly flight behavior). Reciprocally, this global transformation of the local is communicated globally again, as illustrated by Dovey's comment when she finds out about the popularity of the project: "'Holy cow,' she said, standing by the sweater bin, peering into the World Wide Web. 'This is happening on your property? It's like, huge. They've got over a thousand Likes on Facebook'" (Kingsolver 300).

This passage demonstrates the significance of communication technologies in the experience of globality and shows how especially social media may serve to emotionally connect people across the globe.[12] While the importance of mediated global experience has been discussed at length by numerous scholars (Giddens, Tomlinson), and there is no doubt that television, the Internet, and telephone communication make "available a range of perspectives on events beyond that of the 'home culture'" (Tomlinson 116), the question remains how this "bring[s] us phenomenologically closer, closing cultural and moral distance and providing a compelling sense of involvement with distant live and events" (Tomlinson 172). A social media platform, such as Facebook, may play a role in finding an answer to this question. Functioning as an interactive media format, it allows the formerly passive viewer to engage with the issue on display. Demonstrating one's support for the butterflies through a "like" or by posting messages in a common forum enables people from across the globe to connect with one another. Moreover, further advanced telecommunication technology such as the smartphone epitomizes the time-space compression (Harvey) prevalent in a globalized world and allows for the immediately felt shortening of distances. When Dellarobia comes across the two knitters up on the mountain by chance, she is not only struck by the technology of the phone itself but also by what the women tell her: "'People are chuffed to bits on it!' Nelda said. 'You should see the messages we get' ... 'Go knitters, stop global madness, we love you.' 'That's from Australia, it came this morning. Here's another, 'Go ladies, green and clean, from Betty in Staten Island'" (Kingsolver 340). Dellarobia is witnessing the first signs of a transnational community of solidarity emerging on the basis of shared risk exposure, made possible by the Internet and modern technologies.[13] As Thomashow explains in his work on the perception of global environmental change, "One's experience of globality entails a series of conceptual leaps which require metaphor, ideology, and cognition, what historian Benedict Anderson describes as 'imaginary linkages.' Forms of mass communication ... catalyze these linkages" (22).

In Kingsolver's novel, the World Wide Web functions as a site of global connectivity as well as of political empowerment. It becomes a powerful tool in quickly spreading and communicating messages, providing the space for a multitude of perspectives. This is highly relevant for an issue such as climate change, which because of its global scale poses difficulties for communication. Thus, when it comes to a "showdown" between Ovid Byron and news reporter Tina Ultner, during which Ovid unmasks the corruptibility of the news industry and its incompetence in representing and reporting on something "intangible" (Kingsolver 367) such as climate change, it is of crucial importance that Dovey uploads the video of this dispute on YouTube. The media platform allows for the information given in the video to be transmitted all across the world. A comment by Ovid's wife Juliet exemplifies these global possibilities: "I saw it before he called me. A friend of ours in Canada forwarded the link" (393). The Internet underscores the space-time compression of globalization, and makes it possible for Juliet to be informed of an event without being present and without having to talk to the people involved.

While Dellarobia is confronted with the mobility of butterflies, people, and information, she herself does not move but stays within local boundaries. However, this does not mean that her space remains domestic, because cultural deterritorialization nevertheless impacts this locality. Furthermore, as Tomlinson points out, "The paradigmatic experience of global modernity for most people—and this is not of course unrelated to the correlation between income and mobility—is that of staying in one place but experiencing the 'dis-placement' that global modernity *brings to them*" (9). This transformation of local experience encompasses an awareness of how cultural belonging is changed by globalization and extends an invitation to consider oneself in a planetary context. Due to her contact with various people from out of state, Dellarobia is offered a range of perspectives that enable her to see beyond the local and to situate herself beyond her territorial confinement. She describes the shift in local experience, "She felt herself looking at things through their eyes sometimes ... It had altered her sense of things, even in this familiar store where she was examining her purchases with some new regard" (Kingsolver 162). This cosmopolitan transformation, however, may not occur without conflict and can be rather unsettling, since it also "represents a loss of the cultural certainty, even of the existential 'comfort' involved in having the world 'out there' presented to us from the still point of an unchallenged national/local perspective" (Tomlinson 116). To imagine oneself in a larger scaled context requires a questioning of one's own place attachment and a re-negotiation of one's own behavioral patterns. Dellarobia not only begins to critically look at her own choices in life, but also increasingly becomes attuned to the interconnections of her life with distant processes. While in a dollar store, she comes across a butterfly-shaped cloth potholder that is made in China and wonders,

> Did they even *have* monarchs in China? She did not know. But somewhere far from here, someone had taken the trouble to get this exactly right. She smoothed it in her hands and pictured a real person, a small woman in a blue paper hairnet seated at a sewing machine. Someone her own size, a mother most likely, working the presser foot up and down to maneuver the careful lines and acute angles of that stitching. Scrolling out a message, whatever it might be. *Get me out of here.* (175)

Once again, the butterfly functions as a synecdoche for a globally connected world. Considering the production process of the knitted butterfly, Dellarobia begins to see her own life paralleled to the life of a Chinese woman and realizes that even though they are in distant places, their plights may be similar.

The butterfly effect ultimately allows for the emergence of what Tomlinson calls "ethical glocalism" (194)—an understanding of an intricate connectedness in a globalized world while refraining from understanding identity as solely defined by immediate locality. This ethical glocalism, not unlike Heise's eco-cosmopolitanism, acknowledges our planetary belonging and thus "embraces a sense of what unites us as human beings, of common risks and possibilities, of mutual responsibilities" (Tomlinson 194).

Flying Off to New Earths

A close reading of Barbara Kingsolver's *Flight Behavior* shows how the novel engages with the global scale of climate change by foregrounding the intricate relationship between humans and non-humans. As Ovid Byron explains in the novel, "Living systems are sensitive to very small changes" (279), and thus only a small increase in CO_2 emissions in the atmosphere already has far-reaching consequences. *Flight Behavior* succeeds in illustrating these far-reaching consequences, shedding light on the multiple layers and large-scale effects of global warming, while at the same time providing insights on the transformation of the local—Dellarobia Turnbow's home—by global forces.

Loss of biodiversity is postulated as one main risk from climate change, but Kingsolver's novel, unlike many other environmentalist narratives of decline, goes beyond an elegiac story template and engages with the issue of extinction in a more nuanced manner.[14] According to Heise,

> Elegiac modes of storytelling are mobilized in the context of extinct and endangered species as a way of reflecting on particular histories of modernization, and in many cases of articulating resistance to the forms this modernization has taken. Such extinction stories function as a means of representing turning points in human cultural histories, in which the loss of a particular species stands in both for a broader sense of the vanishing of nature and the weakening of human bonds to the natural world. ("Cultures of Extinction" 68–9)

While "declensionist" moments can also be found in *Flight Behavior*, and while the disturbed butterfly migratory behavior certainly symbolizes a larger disruption in the ecological system due to man-made climate change, Kingsolver's placing the butterflies at the verge of extinction paradoxically also serves to convey to the reader the agentic power of nature, which interacts with and relates to human existence. In this context, the narrative refrains from a conservationist agenda proposing a human intervention to save the doomed species. Even though a rethinking of humans' interaction with the environment is necessary, this must take place holistically. As Dellarobia understands through conservations with Ovid Byron, what makes a monarch a monarch is its "interactions with other monarchs, habitat, the migration, everything. The population functions as a whole being" (Kingsolver 317). Thus realizing that the butterfly problem is not only one of physical displacement—just as her own situation cannot be changed by geographical displacement—she comprehends that Ovid is "not a zookeeper" (320) providing the managerial fix to save the butterflies.

The novel challenges elegiac cultural narratives of decline and critically engages with a question that Dellarobia also begins to contemplate—namely, "to whom did a species belong?" (320). *Flight Behavior* thereby complicates notions of villains or victims and avoids moralizing, dualistic environmental rhetoric. Instead, it aims for an awareness of our own situatedness within a "complicated system" (Kingsolver 143), calling for a "planetary consciousness" (Heise) which acknowledges that in the climatically changed world of tomorrow we are not experiencing the "end of nature" but possibly need to understand nature anew.[15]

Even though the final pages of the narrative invoke the biblical deluge of Noah and depict Dellarobia engulfed by a major flood, Kingsolver's story does not propose an apocalyptic ending of the world. Instead, the protagonist is able to save herself from the waters running down her property, and describes in an epiphanic moment how some of the butterflies are flying off to a new place—"a new earth" (433):

> Their numbers astonished her. Maybe a million. The shards of a wrecked generation had rested alive like a heartbeat in trees, snow-covered, charged with resistance. Now the sun blinked open on a long impossible time, and here was the exodus. *They would gather on other fields and risk other odds, probably no better or worse than hers.* (433, emphasis added)

Avoiding the tragic moment of extinction, the novel postulates that life on earth will continue but this life will be "different from the one that has always supported them" and different "in the manner to which we have all grown accustomed" (325). There will be no way of returning to a pre–climate change state, neither to a risk-free environment but there may be ways of, nevertheless, shaping the future differently: The butterfly exodus ultimately resembles Dellarobia's own way to emancipation and empowerment. In his analysis of *Flight Behavior*, also literary critic Axel Goodbody notes that the story and particular its ending move away from apocalyptic rhetoric and instead resemble more a "narrative of an individual's awakening to environmental risk and simultaneous realization of their potential as an active member of society" ("Risk, Denial, and Narrative Form" 48). As Dellarobia invents herself anew—leaving her unhappy marriage, going to university and embracing a new life—and the butterflies are flying off to find new grounds, geographical as well as sociocultural borders are crossed and new habitats created. In *Flight Behavior*, it is then not too late to envision better and more sustainable "new earths."

From Roots to Routes: Riskscapes in Steven Amsterdam's *Things We Didn't See Coming*

While Kingsolver's novel depicts a near-future setting with a stark resemblance to our own and postulates that a planetary sense of belonging may be key to avoid the worst of the climate crisis, Amsterdam's *Things We Didn't See Coming* provides a more dystopian vision, in which "new earths" are as uninhabitable as the old ones. As analyzed in the following, the global dimension of climate change is invoked here by portraying the world as a global riskscape.

While policy measures to reduce CO_2 emissions are being stalled, world temperatures continue to rise. The consequences of such warming are manifold, but as sudden or gradual changes to local environments (such as extreme weather events, but also sea level rise, or desertification) increasingly force people to leave their home countries, the term "climate refugees" gains in prominence and urgency. Facing the transformation of places across the earth consequently challenges us with questions of place-value, attachment to home, and adaptation to mobility. Originating at a global scale, climate risk scenarios intensify deterritorializing processes, "as they prompt individuals and communities to reconfigure their practices of inhabitation in relation to these larger sociospatial scales" (Heise, *Sense of Planet* 152). As a result, this risk perception may lead to a dislodging from place after one's conscious or subconscious assessment of the dangers that this place poses (ibid.).

Turning to Steven Amsterdam's *Things We Didn't See Coming* is informative in this context, not only because it provides an innovative narrative format for representing the deterritorializing aspects of climate risks but also because it deals with the ways global climate risks challenge attachments to places and question forms of belonging and inhabitation. Though localities and "real places" exist in the novel as descriptive settings, they are no longer recognizable as an "'anthropological place'—one that provides cultural identity and memory, binding its inhabitants to the history of the locale through the daily repetitions of 'organic' social interactions" (Tomlinson 109).[16] As a result, places in *Things We Didn't See Coming* are transformed into riskscapes, in which socio-political territorial distinctions have lost in importance and individual attachments to places have been severed due to ever new emerging risks—an aspect which is highlighted in the novel by the namelessness of places and most of the characters.

As the following analysis will demonstrate, Amsterdam's novel thus depicts the climatically changed world of tomorrow as what I call—following Arjun Appadurai's work on globalization—a "riskscape."[17] In "Disjuncture and Difference," Appadurai distinguishes between five different "cultural flows" representing the deterritorialized state of the world: ethnoscapes, finanscapes, mediascapes, technoscapes, and ideoscapes.[18] He uses the notion of *-scapes* not only to describe how globalization has transformed and shaped our experience of time and space, but also to show how these five dimensions are interconnected and continuously in flux (Appadurai 296). Reading Appadurai's work along Ulrich Beck's elaborations on the "risk society," one could thus add the dimension of the "riskscape" to Appadurai's framework.[19]

Accordingly, riskscapes are characterized by the anticipation and experience of a ubiquitous crisis, in which territorial distinctions decline in importance and socio-cultural practices are disembedded from place. Applying the term "riskscape" to Amsterdam's *Things We Didn't See Coming* effectively captures the volatility and constantly shifting meanings of places in the novel. Furthermore, it powerfully demonstrates that climate risks are not geographically confined. The characters in Amsterdam's novel inhabit hazardous riskscapes, and, as a result, are forced into continuous movement. Since these riskscapes impede any form of re-inhabitation, adaptability and constant renegotiation of the self become key to survival.

Riskscapes also resonate with what anthropologist Marc Augé has termed "non-places." He explains, "If a place can be defined as relational, historical and concerned with identity, then a space which cannot be defined as relational or historical or concerned with identity will be a non-place" (77–8). While for Augé these non-places in "supermodernity" are created for a particular purpose—as his examples of the airport, the shopping mall, or highways demonstrate—, they remain spaces of transit and are thus deprived of meaningful communicative practices or relations.[20] Furthermore, he argues that a world dominated by non-places has "surrendered to solitary individuality, to the fleeting, the temporary and ephemeral" (78). Reigned by actuality, demarcated by ideograms and instructions (such as "take right-hand lanes"), non-places are entered and exited but never dwelled in. A key figure in Augé's discussion of non-places is thus the passenger—the person who moves through the world of transit places. Entering these spaces of anonymity, she/he never stays long enough to fill the place with routines or social interactions. Instead, every encounter is superficial, converting these non-places into "places of solitude (even in the presence of others), silence, anonymity, alienation and impermanence" (Tomlinson 110), providing no comfort to the passenger.[21]

Amsterdam's novel illustrates how supermodernity's unintended consequences have led to the creation of riskscapes in the future world affected by climate change. In these riskscapes, non-places are no longer confined to particular places (e.g., the airport), but are omnipresent and impede any dwelling at all. The figure of the passenger, who in Augé's work is defined by his final destination and able to freely enter and exit the non-places he passes through, is no longer able to navigate freely while moving through climatic riskscapes in *Things We Didn't See Coming*. He is thus turned into an eternal passenger.

Textual Manifestations of Global Risks

The title of Amsterdam's novel already alludes to the overarching theme of riskscapes that influences narrative structure and content alike: *Things We Didn't See Coming*. The book offers glimpses into the life of a nameless protagonist and his struggle for survival in a world that is characterized by weather extremes fluctuating between heavy precipitation and severe periods of drought, battles over scarce resources, and an increasing uninhabitability of places. These changing climatic conditions, indicated to be anthropogenic in nature, force the protagonist as well as the other characters in the novel to continuously re-evaluate their living situations. Through nine individual stories, the reader travels in time, starting at the turn of the millennium when the homodiegetic narrator is still a young boy and as the chapter title indicates at "What We Know Now" and ending with the closing of his eyes and thus presumably his death about 40 years later. Since Amsterdam's work follows a stop-start narrative pattern, the nine chapters of the novel are disconnected and could be read independently of one another.[22] Rather than weaving a linear storyline of Aristotelian fashion, chronologically presenting set-up, conflict and resolution, every chapter elaborates a novel idea in a new setting, involving different characters. Hence, just as the protagonist is continuously confronted by newly arising environmental disasters throughout the story, which make it impossible for him to plan his next steps, neither the reader is able to anticipate setting or plot of the next chapter. Readers' expectations are thus constantly questioned, never letting her or him be certain what the next turn in the storyline will bring, where the narrator will be and with whom. Risks then seemingly disrupt the narrative as much as they disrupt the characters' lives. Because of these continuous displacements, the book requires an attentive reader, who with every chapter anew, is forced to realize what she/he did not see coming before. However, as collage-like and self-directed as the chapters are, they nevertheless are connected—on the micro level because if read chronologically they form the life story of the narrator, and on the macro level because as a sum they represent a world transformed by climate change—or as Amsterdam writes in his final chapter, "the itinerary you all signed up for" (185).

Amsterdam's narrative technique is innovative in rendering risk scenarios intelligible as it neither relies on the dominant discourse models as defined by Buell nor on the genre models that Heise discusses for risk narratives, but rather blurs generic boundaries.[23] Though employing elements of the coming-of-age story and resonating elements of the picaresque novel, the narration of *Things We Didn't See Coming* is disrupted to such an extent that generic closure is never really provided. Contrary to the picaresque novel, in which the picaro's heroic act ultimately lies in his resistance against the hypocrisies of the society he lives in, Amsterdam's climate change fiction does not easily give in into this binary good vs. bad situation. This also distinguishes the novel from apocalyptic narratives.

As a recurring theme in environmental discourse, apocalypse—the final destruction of the world—is employed to call attention to the urgency of environmental issues. Heise acknowledges that secular forms of apocalyptic discourse may be understood as a form of risk narrative, but also points to the nuances which distinguish the two: whereas apocalyptic scenarios often posit a global disaster situation in which destruction may be averted in the hope of restoring a pastoral ideal, risk scenarios emphasize "uncertainties, unintended consequences, and necessary trade-offs" (141). Moreover, they deny the possibility of returning to a world without hazards. Amsterdam's novel, similarly to most of climate change fiction and also to Kingsolver's novel, thus can be categorized as risk narrative since the key to survival in its manifold riskscapes is adaptability to a constantly changing environment. Each chapter then focuses on the protagonist's careful considerations of what risks to take and which ones to avoid, as it is evident that the riskscapes he is moving through are not temporary but as protagonist points out "it's permanent" (Amsterdam 101). Turning to Frederick Buell's discussion in *From Apocalypse to Way of Life* in this context is insightful. As explained earlier, Buell argues that environmental discourse has shifted away from apocalypse as "crisis has become domesticated into daily life" (186). Amsterdam's novel underlines this assumption and exemplifies this "slow process of increasing ecological and ecosocial immiseration and rising ecological and ecosocial risk" (ibid.).

In order to survive in these riskscapes, the novel suggests that even immoral or cruel actions are necessary trade-offs. When the narrator encounters two women (mother and daughter) in an abandoned house during his evacuation job, he considers what to do with them, "if Jenna leaves, Liz will wander outside and be dead of exposure within two days, guaranteed. Then I can come back for the wine. *It's not a heartless plan, it saves one of them, at least*" (Amsterdam 60, emphasis added). Complicating the distinction between good and evil, right or wrong in the riskscapes created by climate change, the protagonist explains, "border clashes, the flu, the weather, and all the migrations they caused—none of it has fostered anything like camaraderie" (Amsterdam 162). As the reader is left to wonder about whom to blame as villain or whom to celebrate as hero, the novel avoids the accusative tone that environmental discourse often reverts to.

Correspondingly, the narrator's antagonists remain as invisible and diffuse as the risks surrounding him. This is especially evident in the chapter "The Profit Motive." While trying to apply for a job with the new administration, the protagonist feels constantly under surveillance but does not know by whom: by the woman who is interviewing him and seemingly testing him or by secret cameras watching him. When trying to figure out the right answer to an interviewer's statement "all our lives are big dark disasters," he describes the dilemma, "I cannot agree (and question the optimism of my potential employer) or disagree (and aggravate my assessor)" (Amsterdam 170). The protagonist has to make his choices without really being sure about its effects, and once again, as no closure is provided by the story also the reader is left in a situation of uncertainty. Instead of proposing an end-of-time scenario, the novel thus depicts future climatic riskscapes in which society, aware of the risk of future catastrophes, increasingly "dwells in crisis." Since there is no means of escape and no clear-cut solution to the given problem, the protagonist in *Things We Didn't See Coming* tries to continuously adapt to—or in Buell's words to domesticate—the riskscapes he moves through, which impacts his experience of space and place.

Narrating the Space and Place of Riskscapes

Amsterdam's novel investigates in detail the deterritorializing process whereby everyday meaning becomes detached from place and from local environments. Places are often understood not only as territorially defined spaces but as "centers of felt value" (Tuan 4) that play a crucial role in people's subject formation. As already alluded to in analysis of *Flight Behavior*, emphasis on locality and place as the source of environmental awareness and self-awareness also frequently take center stage in ecocritical discourse. However, this presupposes a rather static understanding of place and neglects the ways in which localities have already been transformed by processes of globalization, thereby becoming increasingly "phantasmagoric" (Giddens), and disembedded from their socio-cultural context. In contrast to the practice of reassurance that Lawrence Buell sees in our tendency to convert "abstract space into familiar place" (*The Environmental Imagination* 261), Amsterdam's novel discusses the effects of what happens if such conversion is no longer possible.

Even though in the opening of the novel, safe places—comfort zones of familiar encounters, rituals, and comforting memories—still exist, they are not carefree spaces since risks are already starting to pervade them. Fearing a major computer server breakdown at the turn of the millennium, the narrator's father urges his family to leave the nameless city he lives in to seek refuge at his grandparents' house in the countryside. However, once the family has arrived, the harmonious atmosphere is disturbed by quarrels over a road accident, causing the family to spend New Year's Eve in different places. While mother and grandparents stay by the stove inside the house, son and father meet up at the "safest place in the world" (Amsterdam 18)—a spot not far away from a hiking trail in which three trees have fallen down into a triangle. Supposedly still protected here by nature, the protagonist's father starts to describe the future disaster scenarios of "breakdowns that can't be fixed" (23). When asked by his young son how to best prepare for such a situation, the father realizes that he does not know the answer. Instead, "he just keeps rocking, telling me he's sorry and hugging me as tight as he can to hold the world still" (24), as the narrator describes. This scene powerfully mirrors how even the supposedly safest place in the world is suddenly pervaded by risks, by the anticipation of a major global breakdown. And even the father, who has allegedly taken all necessary precautions, is not prepared enough "for everything to fall from interdependence" (12), nor is he able to explain how to prepare for the changes and disruptions to come. This introductory chapter illustrates that risks already constitute a dominant aspect of the protagonist's life from the beginning on. Moreover, it alludes to the problem of imagining protective action against global risks. As the father comes to realize, no place—especially not this countryside retreat—can keep his son safe from the global disasters to come. While chapter one of Amsterdam's novel describes an awareness of risks that are more technological in nature, in contrast to later chapters that deal with climate risks, it already suggests the multiple reverberations and global scale that risks may have and challenges the very notion of safe places.

The second chapter of the book, entitled "The theft that got me here" is set in the near future when climate change is already having a noticeable impact on the protagonist's young adult life. The Central's, an unspecified government institution, first reaction to changing climatic conditions has been the division of human living space into urban and rural areas. While here space still is clearly demarcated, with the further lecture of the book the reader is provoked to re-consider this division because concepts such as city or country come to be repeatedly questioned. This is most notable in chapter six when Juliet, one of the political leaders of one fourth of the country, presents her utopia: "Her goal, she says, is ... [that] one day there will be no more cities—their shells will be ghostly interruptions of the new nation, which will be composed of rural communities linked in all directions" (125). In chapter two however, "everybody [still] had to choose, urban or rural" and thus suburbia, for example, has become an empty, "dry" and "decaying" (Amsterdam 33) space. While barricades and special IDs territorially and politically define the place the protagonist inhabits, the city nevertheless remains an empty signifier, because it is deprived of the socio-cultural practices reviving it. As the protagonist explains, the city's streets remain empty and dusty and without any interaction since "no one's exactly getting dressed up for a Sunday drive" (26) or goes out at all. In this environment humans become the mirror of the environments they inhabit: shadow figures, like the narrator's grandmother who has been "off the map for six years," i.e. mentally absent due to Alzheimer, or the protagonist himself who "is lying there decimated, slowed down to a piss puddle of [his] former self" (27), because his only means of dealing with the situation is to sedate himself with drugs.

These deterritorializing tendencies are challenged when all of a sudden the narrator's grandmother is lucid again and attempts to re-territorialize place. Deciding to go on a Sunday drive to the countryside, the family has to face political reality as it is stopped at the barricade demarcating city territory. Yet, the grandmother successfully convinces the patrol officer to let them pass by ultimately restoring meaning to the "countryside" that lies behind the barricade: In vivid terms, she describes not only her former home and lifestyle "out there," thereby reterritorializing the place, but also frees the city of its usually socially sterile atmosphere by drawing a personal relationship between herself and the officer. She points out: "Anything named Central doesn't even know what you look like. I do, and I'm watching you to see that you make the right decision here" (33). The officer then waves the family through.

However, the countryside they encounter does not resemble the pastoral idyll of their memories. They face hostility and several attacks by country-dwellers. After managing to escape the offenders, the family finally comes to rest at a calm picnic spot. Yet, the narrator's description of his grandparents as "happy ghosts" (38) in that moment already foreshadows the following set of events. Just when the grandparents decide to part ways with the protagonist in order to continue their lives in the countryside, the illness takes hold of grandmother again. As a consequence and knowing there will not be another chance of self-determined life, the grandfather administers himself and his wife a deadly pill cocktail. Deprived of his family and without any other social attachments, the lonesome narrator finally decides to move, head west, toward the "endless land" (47). His desire to escape to the wild and "untouched" West and to live a life "on the road" echo key topoi of American culture, which has been elaborated on by Ursula Heise. She explains, "Americans have seen themselves as modern nomads, and have always felt ambivalent about their mobility, perceiving it by turns as their greatest social asset and their deepest cultural deficiency" (49). This conflict is also foregrounded in Amsterdam's novel. As the story proceeds and the narrator moves west, he increasingly finds himself caught in climate change–driven riskscapes, which impede any form of "settlement" and thus force him to be "on the move." His mobility and uprootedness are then permanent, worsening the protagonist's solitude and sense of alienation. The hostility of riskscapes thus revokes the positive notion of the "modern nomad," who is no longer a traveler by choice.

The third chapter, entitled "Dry Land," set at a later point in the future of the story, describes these riskscapes that turn nomads into refugees, travelers into eternal passengers in greater detail. It paradoxically depicts how the land is anything but dry, because climate change has led to increased rain patterns. The reader discovers that the narrator now works as an evacuation officer for "Land Management," looking for "the shaky light of candles burning in dark houses and evacuat[ing] whoever's still thinking the sky's about to clear" (Amsterdam 48–9). While his only companion, a "rain horse," has "been sensitized to travel in downpours without complaint" (48), for him it is more difficult to travel in constant rain but government medication helps to endure these wet conditions. The protagonist describes the transformation of the places he encounters on his evacuation route: the solar sheets which collect "nothing but water now" (51), and the "thousand-dollar appliances you can't use since the grid went down" (52) in a once-precious summer house which now stands abandoned and uninviting. Also, the surrounding landscape is no longer what it used to be, but the "grass or whatever was here before has been stripped by a thousand new rivers coming down from the mountain" (51), ultimately turning this piece of land into an "area for wildlife now, not people" (60). Stressing the volatility of riskscapes, this passage demonstrates how places and landscapes constantly come to be reconfigured.

While living conditions in these changeable riskscapes have become extremely difficult and dangerous, many of the people that the protagonist encounters accept their life in misery instead of realistically assessing the hazards they confront. Similar to Frederic Buell's discussion of contemporary society's domestication of crisis, the characters in Amsterdam's novel thus try to accommodate to life in riskscapes. However, as the narrator explains, this state of "dwelling in crisis" needs to be overcome by one's acceptance of the deterritorializing processes of climate risks and thus by embracing change. This is most prominently illustrated by the narrator's comment about his job. He complains,

> It's the people part of the job that isn't so great ... Some guys who come out on these gigs are full of compassion for the human suffering and want to help the families *come to terms, deal with their loss, adjust.* It's a noble calling, if you see it like that, but all I see is people being washed away by life and I think the respectful thing to do is treat them efficiently, not ask them *how it feels.* (49)

For the protagonist, dealing with this climatic risk situation is a matter of moving on and not one of dwelling on the life and place one has been attached to. After all, in a climatically changed world, those lives and places will sooner or later be "washed away" along with everything else. To continue dwelling in a place in the face of further risks or to develop any emotional personal attachments is futile. Instead, as it becomes evident that landscapes, places, and lives all undergo constant transformations, survival depends on becoming as mobile and independent as possible.

The hostility of riskscapes is mirrored in the personal interactions of the characters. Returning to the passage, in which the narrator meets a woman and her daughter in one of the abandoned houses serves to further illustrate this, since their contact ultimately revolves around the question of survival at the cost of betraying each other. The protagonist quickly decides that both the alcoholic mother "whose body feels like life's gone from it" (66) and the daughter "that's got no instinct to make it" (58) are a lost cause. He thinks of their attachment to one another and their clinging to the past as particularly counterproductive to adaptability, which to him is crucial for survival. Eventually, it is the narrator who is deceived by mother and daughter and forced to leave them and the house behind. He starts to contemplate "what it's going to take to make them give up on each other, turn them into survivors" (69). Startled by a deer passing by, he further wonders, "what these animals were holding on for, before they finally decided to save themselves" (69). The protagonist thus muses on the value of personal as well as place-related attachments and ponders at what point risks become life-threatening enough to break these bonds. Only at a later point in the novel, however, does he give a partial answer to this question: "if you lose everything once, running becomes a part of you" (101). Accordingly, only once people have experienced complete destruction and catastrophe will they take their vulnerability to future risks seriously. They then begin to realize that survival in riskscapes depends on detachment from place and even solitude.

Despite the fact that in the two next chapters the narrator is no longer on his own but accompanied by a woman named Margo, isolation and uprootedness remain prominent themes. When quarantine is declared in the city (again nameless and not further defined), the couple moves out into woods and builds a camp. Margo, who has been restless already in the city and disappeared sometimes for several days—often to go stealing—does not stop her wandering out there, as the narrator hoped she would. He explains his pain, "it hurts to think what she really wants is solitude. I liked it better when I thought she was after more material comforts" (75). His admission reveals how important personal attachment really is to him and that his isolation is a painful state that has been forced upon him as a result of trying to survive in an unpredictable riskscape. However, he soon is reminded of the advantages of a life in isolation. When their camp is invaded by a deadly infected "barely surviving humanist in an inhuman world" (81) as the intruder describes himself, the narrator is no longer only confronted with the fear of not finding enough food or water or with the question of when Margo will return but faces the risk of being infected as well. He paints a vivid image of the scene: "He's on all fours, throwing up blood under the tree. Such a pastoral scene ... the pale late-afternoon sky, the dried green of the trees, the little man bent over like a dog on the ground beneath it all, and the splash of red at the bottom of the canvas, seeping into the soil" (73). The blood figuratively not only breaks the pastoral idyll of the image, but it also literally contaminates the soil and thus ultimately the place which he inhabits and which served as the basis for his survival. The risk of being infected then functions as yet another deterritorializing moment, forcing the protagonist to abandon this place.

Upon Margo's return from one of her wanderings to the contaminated site, they thus immediately begin to look for a new place to install themselves. Coming to an abandoned campsite, the protagonist feels the urge to "move in and make their world mine" (89). However, he anticipates that in doing so "I am *sure in my blood* we'll doom ourselves to always live exactly as we have lived, inhabiting whatever corner of the world isn't nailed down, never staying anywhere long enough to make anything real. We will be the ghosts that feed off the edges of life" (Amsterdam 89). Interestingly, at this point the protagonist fears that their nomadic life will never come to an end and is overwhelmed by the yearning that attachment to one particular place is still possible and will enable them to "make something real." Yet, when the other residents return unexpectedly, "the decision is made for [them]" (ibid.) and settling at the campsite is no longer an option. At this point of the narration, all hopes for the possibility of re-inhabitation are thus shattered and the narrator's mantra is introduced, which is to always move on and to hurry "if we're going to make it" (109). Consequently, he and Margo are never staying long enough in one place to enliven it with socio-cultural practices because they are only waiting "for the next event to push [them] along" (104). Deprived of meaning, places are no longer reference points, but, following Augé's discussion, transit points or non-places. Wondering where to go without knowing where they are, Margo and the protagonist become the eternal passengers that the protagonist was afraid they would turn into.

As the story continues, it becomes apparent that it is not just the narrator and his partner who are forced to continuously be "on the move," but everyone else also is. Due to increasingly extreme weather events, places have been rated according to a "Most Livable List" (Amsterdam 93) and the narrator now works as a distributer of cash grants for relocation in places that have to be abandoned. No longer dealing with heavy rain but extreme heat, droughts and mosquito attacks, the scene is described as devastating, leaving staff members no choice but to take pills to cope with the situation. The few inhabitants who still remain at disasters sites are "laughing about their dumb luck for surviving. But they have this newborn worry in their faces" (101), because though "the people in Grief work hard to push the line that rebuilding heals all … nothing really heals because … you're always looking back" (101). They already suspect that survival in riskscapes will based on constant relocation, nostalgically looking back at what has been lost and longing for earlier and more stable times. The protagonist also suffers from the situation of constant displacement, but at this point of the book his longing for stability is mostly suppressed and only subconsciously emerges in nightmares about a young boy wanting to be led to his home, or it materializes in panic attacks when he is overwhelmed by the fear of the future (Amsterdam 95, 103). Aware that a return to a risk-free environment will not be possible, he makes fun of the other evacuees who still believe in the volatility of the situation:

> Most evacuees don't learn. They try to start over someplace exciting (a target) or temperate (subjects to floods, fires, or earthquakes). Or they identify the month's most thermopolitically neutral region. They assume they're not going to have to pack again. Even though it may be the third or fourth time for some of them, they're still completely tweaked with relocation fever … They take their first steps around their new home and get confident; make friends, buy appliances, plant tomatoes. You want to shake them: *Do you really think this time it's going to be different?* (Amsterdam 104)

This passage demonstrates that though there still remains an urge for reterritorialization and a need to seek psychological comfort in the sociocultural attachment to place, this is ultimately futile. Riskscapes no longer allow for re-inhabitation. Juxtaposing rootedness and mobility, the narrator clearly indicates that only the latter can serve as the "route to survival." Humans then literally become "natural aliens," a species without fixed habitat, that environmental studies scholar Neil Evernden has described. This is especially prominent in a moment of global crisis. Evernden tellingly explains, "in anxiety the world falls away and loses all significance, so that one is unattached, at home nowhere" (120), and in Amsterdam's case lost as an eternal passenger in meaningless spaces of transit.

The chapter "The Forest for the Trees" critically envisions this placelessness and effectively demonstrates the difficulties between the negotiation of risks, uprootedness, and eternal mobility on the one hand and security, emplacement, and stability on the other. In what seemingly are fairly stable times, the narrator works as a speechwriter for a woman named Juliet, the political leader of a fourth of the population of the country, while also leading a very intimate relationship with both her and Margo. However, pressured to conform to Juliet's wishes, he and Margo, as the narrator explains, are always pretending to be on their best behavior. Returning to Augé's discussion helps to illustrate how the passenger always performs identity ("shells") anew in non-places. Augé explains, "subjected to a gentle form of possession, to which he surrenders himself with more or less talent or conviction—he tastes for a while like anyone who is possessed—the passive joys of identity-loss, and the more active pleasure of role-playing" (103). Indeed, for a while the protagonist enjoys playing the new role that has been ascribed to him, especially because all records of his and Margo's criminal activities in the past have been erased. However, the narrator is aware that this is only a performance and he hopes that by being more thoughtful and better intentioned "[we] someday ... will be who we really are" (Amsterdam 120), invoking again the need to "make something real." When the time comes to renew his "practical union" with Margo, the narrator also sees it as an opportunity to secure their relationship with Juliet, a person known to let "people go as easily as she has taken [them] on" (121), by adding Juliet to their practical union contract. He is tired of not knowing what the future will bring and of thus being continuously prepared to change again. This is illustrated most explicitly when he explains his plan to Margo: "Simple security. Wouldn't it be nice to know where we're going to be next year?" (123). According to the protagonist, a three-way union would not only benefit him and Margo by providing them with stability, but it would also help Juliet, who is often questioned by the public about when she would finally settle down. The union would credit her with a much-needed "uniter" (125) image, mirroring her own endeavors to politically and territorially unite the country.

The narrator's attempts to reduce his exposure to risks by striving for more security in climatically and politically unstable surroundings, however, become absurd toward the end of the chapter, when his plan literally goes up in smoke. Forced by Juliet to participate in one of her power games, his urge for longer-term security is dismissed by his own short-term risk-taking, as she convinces him to put on a supposedly fireproof suit in order to set a forest on fire. Though Juliet and Margo assure him that "everyone, everything is safe" (135), the flames of the forest surrounding him stand in stark contrast to this assertion. Upon Margo's fatalism "Light one [branch] and they all go," the protagonist realizes that ultimately his relationship with her has never been comforting but that "as long as [he] ha[s] known her, [he] ha[s] never known peace" (134). Also his plan to gain security by a three-way union with Juliet and Margo has been an illusion. His final thoughts, "I no longer want anything at all" (136), suggest his resignation as he looks at the fire quickly spreading around him and finally understands holding on to a person, an idea, or even a place is pointless because no matter where he will be, his environment will never be free from risk.

Finally, toward the end of the book not only the environment, but also the human body itself is revealed as a riskscape. Even though the protagonist believes himself to be healthy, he discovers a medical report which indicates that his body has several "ticking bombs inside" (148), with symptoms in the nervous, lymphatic, and digestive system expected to manifest over the next several years. As riskscapes function as a stressor to human health conditions, the inner space (the human body) increasingly figures as the broken mirror of the non-places it inhabits—something that has already been foreshadowed by the infected intruder in the novel's campsite scene. Thus, body and environment alike pose significant threats to the characters in the novel. Apart from the physiological diagnosis, the medical report, moreover, gives insight into the protagonist's social and psychiatric condition:

> [...] strong cognition and problem-solving, weak emotive-spiritual; tendency toward risk-taking/impatience/dishonesty when stressed; formative years yielded strong desire/ability to trust, make attachments—countered by apparent experience; optimism subnormal, conflicts with strong survival instinct; isolationist, dysthymic tendencies expected to increase with age. (147)

This risk account then functions as a short summary of the narrator's psychological development so far and reinforces his need for attachment. This need, however, has remained unanswered by the outer riskscapes he has moved through. Ultimately, uprootedness and continuous displacement not only have had an effect on his psychological constitution, but have also led to bodily suffering. Again, being put in a situation of risk evaluation, the narrator follows his "survival instincts," thereby confirming what the report has said about him: he betrays the community he currently lives with. Even though he "really wanted to call this place home" (149), he breaks the rules and leaves the group behind in order to take his body "full of problems" (147) to a medical complex. After a quick and efficient medical treatment, he vacillates between returning to the community he left and taking care of the young boy that came with him to the treatment center. Finally, he decides against both options and chooses the road instead. He explains, "it feels good, it's where I belong: I've escaped" (161).

However, this escape is only temporary, since neither the roads he travels on nor the medical treatments he receives will provide security. Instead, as the final chapter of the novel reveals, the narrator's body is no longer able to bear the riskscapes of the future, climatically changed world and finally breaks down. Working as a tour guide for groups of people with "at least two major cancers or a primary ailment" (184), who have been through many ineffective medical treatments and now just want to experience some thrill (volcano outbreaks, safaris) before they die, the protagonist realizes that he himself looks "like the rest of them" (194). He himself has become part of the terminally ill. Not only is his skin disintegrating, but in the final encounter with his father (who now lives as a healer in the countryside), the protagonist lets go of the last string of life he is still holding on to, and "without the slightest incantation, [the father] closes [his] eyes" (199). Circling back to the beginning and the moment in which the first decisions in the face of risk were taken, the novel then seems to repeat the narrator's question: Since crisis is inevitable, how can one best prepare for the volatility of riskscapes and changing environments—especially if crisis forces one to live without "homes" or communities of belonging? And are we psychologically and physically equipped to adapt to a life as passengers in non-places?

Amsterdam's *Things We Didn't See Coming* zooms in on the personal experience and puts a human face on this otherwise "disembodied idea," while at the same time giving credit to the deterritorializing aspects of global climate change. It thus figures as an innovative response to climate change as a "disembodied and unsituated" phenomenon "with its universalizing demands on our imaginations and behaviours" (Hulme, "Cosmopolitan Climates" 273). As the novel illustrates, deterritorialization does not mean that place is completely annihilated, because, as Tomlinson points out, "we are all, as human beings, embodied and physically located" (149) and the need to re-territorialize continues to exercise significant power over our behavior and psychology. However, what global risks such as climate change confront us with is that as places are continuously transformed, ties to them are significantly weakened. Amsterdam's work thus challenges our understanding of place-attachment because it illustrates that in the future climatically changed world humans will move in riskscapes, in which uprootedness will become the norm.

Ignoring issues of social inequality that climate change related deterritorialization introduces, the novel's emphasis on nameless places and mostly nameless characters seems to suggest that in global riskscapes, it longer matters where we come from or who we are and thus reflects the notion that these events may happen anywhere and to anyone. As place is transformed into non-place through the deterritorializing aspects of climate risks, everyone is turned into an eternal migrant. For the characters' survival, these hostile and ever-changing environments mean constant adaptation and a life of isolation and solitude. Thus understood, deterritorialization and mobility do not enable a cosmopolitan thinking—described by Ulrich Beck as an emergence of a global moral consciousness based on shared risk experience (*World Risk Society* 14–15)—but rather create solitary passengers as the tragic unintended consequences of today's risk society. In contrast to Kingsolver's novel, Amsterdam's vision of a "planetary consciousness" then involves the awareness of a global uprootedness and not of a global connectedness.

Amsterdam's novel is significant to the growing body of climate change fiction because it engages with the representational challenge that global risks pose in novel ways. Through its fragmented story-telling, it puts characters and readers alike in a position where they cannot know what is coming, which effectively illustrates that although one cannot fully "know" risks, one nevertheless has to make decisions in the face of them—i.e., the characters have to decide whether to change location and readers whether to continue reading. The novel furthermore innovatively deals with the deterritorialized and global dimension of climate change risks without having to rely on digital networks or other means of disembodied, virtual communication, which Heise considers fitting for the representation of large-scale phenomena such as climate change (*Sense of Planet* 209). In her analysis of climate change fiction, Heise points out that most literary representations fall short of illustrating the magnitude of such a planetary transformation and of envisioning "less territorially defined forms of inhabitation" (207). David Brin's *Earth* is credited as the only successful example of engaging with a global environmental understanding. Heise reads Brin's fractured and postmodern writing techniques as well as the novel's prominent commitment to digital connectedness as characteristics of a commitment to a multivocal, multicultural perspective and thus as an effective means of narrativizing *global* risks (208). However, critically examining her line of thought, Mike Ziser notes, "understanding climate change and peak oil requires us to acknowledge the ways in which our best critical paradigms, avant- and derriere-garde alike, reflect rather than contest our overconsumption of petroleum and the overemission of carbon dioxide and other greenhouse gases" ("Home Again" 187). Ziser's observation is helpful here because it alludes to an envisioning of the future, similar to Amsterdam's, which acknowledges possible material limitations. Ziser's argument thus may serve as an explanation as to why technological connectivity does not play a prominent role in some climate change fictions. This, however, does not mean that the global scale of climate change is not represented. On the contrary, Amsterdam's *Things We Didn't See Coming* provides an original narrative template for the representation of global risks because it refuses to focus on one particular locality or place. Instead, the book painfully relates to the reader that if she or he were to zoom in on place in a climatically changed world of the future, that place would no longer be recognizable because it would have become part of ever-shifting global riskscapes.

The ways of engaging with *global* climate change are manifold as the two analyses in this chapter demonstrate. Yet, it is important to note that both, Amsterdam's and Kingsolver's book, point to the importance of a "planetary consciousness" in trying to understand the global climate change crisis. *Flight Behavior* as well as *Things We Didn't See Coming* thus invite readers to think beyond territorial boundaries and aim to foster a "sense of planet." Global climate change risks in these books lead not only more physical mobility, but, more importantly, also cause the transformation of the places of those who stay put. Showing how locality and place become deterritorialized, Amsterdam and Kingsolver employ novel yet very different narrative strategies. Amsterdam's dystopian vision builds on global uprootedness and tellingly portrays how place is deprived of meaning if everyone becomes a "climate refugee." In Kingsolver's work, on the other hand, the transformation of the local by global climate risks is not yet as severe. Her novel thus presents a more utopian vision, which leaves room for hope that a global environmental consciousness may be key to open up paths to more sustainable futures. Nevertheless, both works in the end seem to agree that a response to the climate crisis requires a letting go of place-based routines and foci.

Notes

1. My translation. Original in German: "Menschen nehmen täglich Wetterereignisse wahr, nicht jedoch das Klima, das in seiner modernen Definition ein statistisch erzeugtes, abstraktes Forschungsobjekt in Langzeitperspektive ist … Zudem können Menschen immer nur einen lokalen Ausschnitt des Wetters erleben, nicht jedoch die globalen Auswirkungen des Klimawandels, was ja voraussetzen würde, nicht weniger als die Welt als Ganze wahrzunehmen."
2. The term "cosmopolitanism" forms part of a series of concepts in globalization studies that attempt to acknowledge the multi-positionality of individuals and places within a globalized world (see Mayer; Pries; Giddens, *Consequences*; Robbins and Cheah). In the first part of this chapter, I will focus on Heise's adaptation of the term "eco-cosmopolitanism" for ecocritical analysis.
3. Drawing here on Benedict Anderson, who with his concept of "imagined communities" describes nations as historically and culturally constructed projections, Heise expands this idea to consider the importance of non-human actors in the formation of communities.
4. Though uprootedness and mobility may be a consequence of risk perception, this does not necessarily imply an emotional detachment from place. On the contrary, as it has been widely discussed in postcolonial discourse (see also Salman Rushdie's *Imaginary Homelands*), an imposed exile often rather intensifies the nostalgic attachment to home and gives in to an elegiac mourning for the home that has been lost.
5. The homophonous pun is Clifford's.
6. Anthropologists such as Kirksey and Helmreich or Doreen Bird Rose, have recently argued for a "multispecies ethnography" which would include those "countless other creatures who are less visible, less beautiful, less a part of our cultural lives" (Rose and Dooren 1).
7. I am aware that using the term "non-human" is problematic in itself due to its anthropocentric perspective.
8. In their essay on "The Emergence of Multispecies Ethnography,“ Kirksey and Helmreich further elaborate on the meaning of *anthropos* for a research approach that aims to go beyond the human-nature chasm (548–549).
9. Gayatri Spivak's term "planetarity" encompasses a rethinking of how communities are perceived. She proposes that instead of seeing ourselves in opposition to the "other," a planetary thinking impedes a

"choosing between cultures" and enables a much-needed transnational literacy (Spivak, "World Systems & the Creole" 108). Spivak has, however, also pointed out that her use of the term does not amount to an applicable methodology ("Response"). For ecocritics, turning to Heise's elaboration on the term in the larger context of "eco-cosmopolitanism" may be more fruitful, because she elaborates in depth on the emergence of a "world citizenship" (*Sense of Planet* 10) that planetarity would entail.

10. In "Rethinking Agrarianism, Place and Citizenship," Janet Fiskio analyzes how the profound connection between land and people can be related back to Thomas Jefferson's republican vision, which postulated that close contact and working the land were central to good citizenship (303).
11. Moreover, Heise has convincingly criticized environmentalist discourse by drawing on postcolonial and cultural studies in order to show how essentialist nation- and place-based identity concepts have failed to recognize the hybrid and performative aspects of identity formation.
12. Tomlinson's discussion of the deterritorializing affects of global technology proves to be insightful here because it exemplifies how the "threshold" determining the constitution of the self may shift in order to encompass more than just the domestic context (Tomlinson 118).
13. This resonates with Beck's idea of a cosmopolitan manifesto. For a further elaboration of this utopian moment in risk societies, see Beck's *World Risk Society;* for its relation to eco-cosmopolitanism, see Heise's *Sense of Place and Sense of Planet.*
14. In "Lost Dogs, Last Birds, and Listed Species: Cultures of Extinction," Ursula Heise elaborates on the tragic and elegiac mode of narrating stories about species extinction in environmental discourse, looking at narratives of decline, such as Quammen's *The Song of the Dodo* (1996).
15. In her analysis of David Brin's *Earth* Heise first introduces the term „planetary consciousness" (*Sense of Planet* 84).
16. Tomlinson summarizes here key aspects of Marc Augé's work and refers to the anthropologist's differentiation between "anthropological places" and "non-places." It is important to note that Augé's "anthropological place" does not imply that this place is "naturally" created, but the author acknowledges that this space is also symbolically constructed and functions on various scales (Augé 51–52).

17. The term "riskscape" was first introduced by geographer Susan Cutter who used it in order to describe the complex interaction between environment, technology and society in the production of specific hazardous landscapes. Pointing out the particular vulnerabilities of a place, Cutter refers with the term riskscapes to very specific and closely defined geographic areas. My employment of the term and reading of it alongside globalization theory, however, aims to go beyond the specificities of place in order to show how climate change creates *global* riskscapes.
18. In order to emphasize that "culture" in a globalized world is not confined to one particular place, but rather takes shape in a multitude of interconnected transnational flows, Appadurai uses the idea of—scapes. Ethnoscapes describe the deterritorialized spaces in which people, often migrants, move and live between different countries. Technoscapes, such as the Internet, allow for global technological connectivity, and stand in close relation to financescapes which designate the global flow of capital. Mediascapes describe the dissemination of information, which construct 'narratives' beyond the nation state and are similar to ideoscapes, which promote political ideas and perspectives about how we perceive the world through images and texts ("Disjuncture and Difference" 296–300).
19. As German sociologist Ulrich Beck explains, the term "risk society," "epitomizes an era of modern society that no longer merely casts off the tradition ways of life but rather wrestles with the side effects of successful modernization—with precarious biographies and inscrutable threats that affect everybody and against which nobody can adequately insure" (Beck, *World at Risk* 8). See also my elaborations in the previous chapter.
20. Augé characterizes the term "supermodernity" with "three figures of excess: overabundance of events, spatial overabundance and the individualization of references" (109).
21. While Augé refers to the person passing through transit places and being determined by his destination as a passenger, Tomlinson points out that Augé's analysis relies in itself on the standpoint of a passenger. Explaining that the perception of places as non-places is dependent on perspective, Tomlinson claims that Augé himself is the "eternal passenger, always viewing locals as one passing through them—on trains, in cars, on planes" (Tomlinson 111).

22. The "stop-start pattern" is a narrative technique used for nonlinear narratives. However, there does not yet exist a fixed definition of the term. In Amsterdam's case, the stop-start pattern emphasizes disruption and the protagonist's own mobility and displacements.
23. In his article "Toxic Discourse" Buell identifies four rhetorical templates in the representation of toxic risks. These include "the disrupted pastoral," "total toxic dispersion," "David vs. Goliath scenario" and "the gothic." Heise furthermore elaborates that risk scenarios often also rely on genre models such as, for example, the detective story, tragedy, or epic (*Sense of Planet* 139).

Reimagining Time in Climate Change Fiction

As living beings we are time, we live time, we feel and perceive time; as human beings we know and reckon time, as members of contemporary Western societies we have externalised time, created it in machine form and now relate to this time as a resource to be sold, allocated and controlled. (Barbara Adam, "Re-vision" 92)

In addition to the challenge of comprehending climate change's global dimensions, understanding this phenomenon requires us to bridge the wide gap between our short-term thinking and decision-making and the resulting long-term environmental consequences. How does one think about something as intangible and invisible as climate change, which does not affect one's life immediately but possibly at some time in the future? And if one cannot notice the effects of it in the present moment, why should one act on them? These questions describe the "Giddens Paradox" or what social psychologists have named "future discounting"—that is, the wide knowledge gap between the familiar preoccupations of everyday life and the abstract future of a climatically changed world (Giddens, *Politics of Climate Change* 2).

The imaginary realm of literature has the potential to allow for the necessary translation of abstract future into individual life stories and the reconnection between actions and impacts.[1] As Jennifer Rose White has explained, literature "can project our understanding and appreciation of invisible, slow, and slowly accelerating crisis into the future in a dramatic

A. Mehnert, *Climate Change Fictions*,
DOI 10.1007/978-3-319-40337-3_4

way that other forms of discourse lack. [Novels] can also legitimately collapse or juxtapose time for maximal impact and understanding in a way that science, biology, and even history cannot" (240). Unlike the temporal representation in scientific diagrams, in which the future indicated on the x-axis spatially marks the difference between the reader's present and his future, climate change fiction deliberately challenges this division: past, present, and future become inseparably intertwined, thereby highlighting that the dangers of tomorrow are already present today. Critically assessing scientific future scenarios, Mike Hulme argues:

> Foresight then—the act of seeing ahead—cannot be limited to the overreaching and hegemonic claims of physical prediction. Instead, a cosmopolitan view of climate change will recognize that our future foresight—and hence our future—is as conditioned by the hopes and fears emerging from the present as it is revealed inside the electronics of a computer model. The climate crisis—if indeed there is a crisis—is a crisis of today even if we would rather depict it as a crisis of tomorrow. *The future and the present are interacting in new ways as we tell ourselves the story of climate change.* ("Cosmopolitan Climates" 171–2, emphasis added)

Though Hulme's argument is central for climate change discourse, because it reveals that future predictions are never independent of value judgments, it lacks an explanation as to how a remaking of the present-future relationship is in fact told, portrayed, and made intelligible.

Moreover, apart from the "technological domestication of time" (Simpson 5), prominently represented by climate modeling and often interested in the future for the purpose of controlling it, contemporary Western societies increasingly focus on the present—a present that is unbound from past or future. Modern technologies, globalization, and ever faster chains of consumption and production have led to a shortening of time horizons to the extent that the present seems to be all there is (Heise, *Chronochisms* 26). Television and the Internet make it possible to bring distant places not only spatially but also temporally closer. The sense of simultaneity is intensified by "just-in-time" logistics, "ready-on-demand" production, and a "throw-away" lifestyle. In this state of a "permanent present," individuals and society as a whole are no longer able to organize past, present, and future into coherent experiences (Jameson, *Postmodernism* 26). As the synchronic is favored over the diachronic perspective, the rhetoric of tipping points, which stresses the disruption of *current* moments in time, also gains prominence. However, this focus on the spectacular and momentary neglects slow and long-term climatic changes.

Moreover, the predominance of the present and the short-term dimension leads to an annihilation of environmental time. Barbara Adam describes environmental time as "latency and immanence, pace and intensity, contingency and context dependence, time-distanciation and intergenerational impacts, rhythmicity and timescales of change, timing and tempo, transience and transcendence, irreversibility and indeterminacy" (*Timescapes of Modernity* 55). Thus, environmental time encompasses those taken-for-granted processes and transformations of nature, which often remain undetected by the human eye. Adam explains that industrialization has had a significant impact on our understanding of time, ultimately aiming at dominating environmental time and making it invisible. She elaborates that three configurations of industrial time—machine, economic, and laboratory time—have turned time into a "quantifiable resource" (*Timescapes* 11).[2] Such an understanding of time enforces the dualistic division between nature and culture, because nature is turned here into solely a material product and considered a means of exploitation. While industrial time is based on an atemporal notion of nature and focuses on visible and tangible products, it simultaneously creates byproducts that are immaterial and last beyond the human timespan. Radiation, climate change, and the effects of toxic waste are only a few examples of environmental processes whose temporal dimension exceeds the temporal dimension of industrial production cycles. Contemporary environmental problems, involving problems of latency, thus require a rethinking of the temporal dimension of nature and a refocusing on environmental time.

In order to "see the hazards of an industrial way of life" (Adam, *Timescapes* 11) nature cannot be discussed solely in terms of place, as often happens in environmental discourse, but also needs to be considered in terms of time. Especially because time has become the "invisible 'other' that works outside and beyond the reach of our senses" (ibid.), we need to refocus our attention on the temporal processes of the environment. With a timescape perspective then, "The invisible becomes tangible and we begin to recognise processes that work below the surface until they materialize as symptoms—sometime, somewhere" (ibid.). However, as already explained above, since today's digital media privilege short attention spans and are always searching for the next spectacular event to report on, incremental or long-term hazardous environmental effects on future generations rarely make headlines.

In his seminal work *Slow Violence and the Environmentalism of the Poor*, Rob Nixon tellingly points to the representational challenges of making slow environmental time reappear. He employs the term "slow violence" in order to describe a kind of violence that is not sensational but occurs gradually, and, for the majority of the "North-Atlantic bourgeoisie," out of sight and in distant places. Nixon stresses the urgent need to direct the public's attention toward these slow and long-term processes and explains that *apprehension*, "a crossover term that draws together the domains of perception, emotion, and action" (14), is crucial to make sense of these otherwise invisible threats. Fictional texts, Nixon says, can serve as a means to foster this state of apprehension: "In a world permeated by insidious, yet unseen or imperceptible violence, imaginative writing can help make the unapparent appear" (15). He stresses that literature proves to be a very suitable realm to explore how slow violence, and thus ultimately also environmental timescapes, are rendered visible. Fiction offers the resources to dramatize and negotiate between the short and the long term, creating narrative relationships between past, present, and future, between humans and the environment while exposing the potential dangers that a bifurcation of time presents. Particularly in the case of climate change, which is so difficult to perceive because of its latency, a more profound engagement with time may open up new perspectives on environmental temporalities.

A close analysis of the representations of time and temporal narrative techniques in climate change fiction thus can serve to illustrate the dynamics of environmental time and reveal the structures of slow violence that go along with the climate change crisis. The works of fiction discussed in this chapter, T.C. Boyle's *A Friend of the Earth* and Jean McNeil's *The Ice Lovers*, explicitly aim to connect present and future, and, thereby, focus on the transformative character of environmental time. They trace and expose the stories that lie beneath the scientific graphs of climate change scenarios, refuting the notion that the future is unbound from the present or past. While taking on the challenge of describing a climatically changed future, Boyle and McNeil simultaneously call attention to the present moment of crisis—the moment in which decisions have to be made—and reconnect past as well as present decision-making with their outcomes in the future.

They are nevertheless aware of the prognostic and thus imaginary dimension inherent in every future projection. Yet, as Helen, the protagonist in McNeil's novel, states, this can also function to the advantage of fiction: "In life we may not be able to know the path not taken, but in books we can" (332). Thus, in both novels the act of writing takes center stage, thereby emphasizing how writers put into words that which cannot be known or seen. Whereas in Boyle's novel writing functions as a means to remember "for posterity" and to record events of the past, in McNeil's novel writing serves more explicitly as a way to imagine alternative futures. Both works ultimately stress that it is important to bridge and reconnect time through narratives and storytelling.

Dealing with the representational challenges of a gradual and thus hardly visible phenomenon such as global warming, the two works discussed in this chapter provide a timescape perspective that engages in the temporal aspects of living in a climatically changing world. In both novels environmental hazards already begin to permeate the present, but become even worse in the readers' near future. While *A Friend of the Earth* aims to provide a perspective that bridges the temporal disconnect, *The Ice Lovers* postulates that the affective mode is important for the experience of time. The following close reading analysis illustrates how both novels articulate the slow and long temporal dimension of climate change and point to the dangers of solely focusing on present-based, short-term thinking.

"Back to the Future"—Bridging Time in T.C. Boyle's *A Friend of the Earth*

In the early summer months of 2013, headlines about severe flooding dominated German newspapers. Some journalists even called it the "hundred-year flood." The flooding along the banks of Elbe, Danube, and several other smaller rivers in Germany was devastating and led to evacuations in cities across the country. It took weeks before the water subsided again. Describing the events at the time, German author Thomas Rosenlöcher notes in his diary, "I am sitting here up high in the Erzgebirge while down in the Elbe valley the waters are still rising" (Rosenlöcher, own translation). However, when scanning through the previous pages of this diary, the writer notices that he had written down almost the same lines only a decade earlier during what was then also coined the "hundred-year flood." Even though the flooding of 2002 was equally devastating, was it possibly not severe enough to be remembered eleven years later? What had prevented journalists from seeing the connection between these past events and present ones? And if it is seemingly already difficult to draw the relation between Germany's "hundred-year flood" in 2002 and its "hundred-year flood" in 2013, one wonders how the more subtle changes that climate change invokes, such as slowly rising sea levels or the increased salinity of the oceans, can be noticed. Our responses to the flooding exemplify what has been elaborated on in the introductory pages to this chapter, that is, our dwelling in the present—a present understood to be unbound by past or future events, oblivious to changes over time.

The idea that contemporary Western societies are increasingly absorbed by the present is underscored by socio-psychological studies on the perception of environmental changes, and in particular by the idea of "shifting baselines." "Shifting baselines," a term originally coined by marine biologist Daniel Pauly, describes one's own changing perception in accordance to the changes of the surrounding environment. So, what ten years ago may have served as a point of reference and criteria of judgment slowly shifts to fit into a new collective reference context, leading to a rather decoupled understanding of environmental changes.[3] To frame this idea more provocatively, it means people are very good at self-deception in order not to be bothered too much by the transformations taking place around them (Leggewie and Welzer 94). In addition to the shifting reference points for climatic changes, Harald Welzer more importantly points to the danger of shifting baselines for the growing injustice induced by climate change. He polemically postulates that what may now still be considered a catastrophe—such as the deaths of so-called "boat people," refugees who drown or die of dehydration before reaching the shores of Europe—may soon be considered the norm by many (Welzer, *Klimakriege* 276–277). While Welzer certainly underestimates human's empathy and sense of injustice and thus takes the idea of shifting baselines to the misanthropic extreme, he effectively demonstrates that active remembering is required in order to counter cultural and environmental amnesia. But what stories and narratives transcend the present and allow for perspectives across time? How can the "dictatorship of the now," as tellingly proclaimed by Hans Joachim Schellnhuber, director of the Potsdam Institute for Climate Impact Research, be overcome?[4] For Schellnhuber, it is obvious that Western societies today need to stop the shortsighted politics of self-deception and should commit to an intergenerational contract that legally anchors the rights of future generations in the constitution. Ultimately, both Welzer and Schellnhuber thus argue that a perspective that incorporates a longer timescape is of great importance when dealing with climate change. They believe that considering the transformations that have taken place over time and trying to imagine what will happen to future generations are crucial in breaking with the dictatorship of the now.

T.C. Boyle's *A Friend of the Earth* similarly warns of the risks involved in such a temporal divide. At the same time, however, the novel aims to bridge that disconnect by offering a narrative framework that allows the reader to reconnect past, present, and future. Boyle thereby challenges the deception caused by shifting baselines, and urges the reader to see the dramatic changes and environmental transformations that occurred between fictional past and fictional future. Furthermore, by directly addressing the reader in his narrative, Boyle reminds the reader of the intergenerational contract and critically questions what kind of a world she/he wants to leave behind.

The opening of *A Friend of the Earth* takes the reader to the year 2025. Climate change has significantly altered living conditions in California: vineyards have been transformed into rice paddies, and since "nobody's insured for weather anymore" (2) constant rain, strong winds, hail storms, and extreme heat waves have left people homeless in Santa Barbara county, Los Andiegoles and San Jose Francisco. With the collapse of the biosphere, many animal and plant species are extinct and humans are living with the constant threat of deadly disease. In this dystopian setting, 75-year-old Ty Tierwater, a former radical environmentalist, works as an "animal-man." This means caring for the private zoo of one-hit wonder rock star, Mac, who considers it a "selfless," "cool," and "brave" task to save those animals that nobody else would have cared for (11). While the story starts out in the year 2025, the novel ultimately delineates two narrative chronologies, which need to be mapped out in order to discuss how the book structures temporal patterns, especially with regard to environmental time. While one storyline follows the events from 2025 to 2026 and is told from the perspective of the autodiegetic narrator, Ty, another strand recounts Ty's radicalization as an environmental activist as well as the series of events that build up to his daughter's death between the years from 1989 to 1997. Notably, a heterodiegetic narrator describes these past events.

Though the overall length of the chapters taking place around 2025 compared to chapters taking place after 1989 is fairly balanced, the narrated time of the past (*histoire*) is much longer than the story in the novel's (future) present. A great part of the narrative set in 2025 is used to describe California after "everything was poached and encroached out of existence" (Boyle 7); however, the focus of the novel seems to be on the past events that have led to this state of devastation. Looking back to July 1989, to "the beginning, the real beginning, of everything to come" (Boyle 24) and then to the eight years that follow, the novel describes, with a great deal of irony, how eco-activists have failed to impede ecological catastrophes. As Richard Kerridge explains, the irony used here, is not "to discredit the environmentalists but merely emphasizes their powerlessness" ("Narratives of Resignation" 90) against a surrounding culture in which their protest is not heard (Boyle 35). Furthermore, the novel's emphasis on the past serves a double purpose: the narrative reconstructs historical coherence in a world that increasingly focuses on the present, and also portrays changes over time. At the same time, it accentuates that books can help restore this temporal continuity, bringing forth memories and knowledge of the past. For instance, the narrator recounts the past events as part of a biography about his daughter, discussed more in detail in the following.

Making Timescapes Visible: Narrating the Past–Present

The novel challenges the loss of perspective, which prevents us from considering the relations between short-term decision-making and long-term environmental consequences, and aims to bridge this temporal disconnect, however painful it may be. As Ty explains, "I try to avoid perspective as much as possible. Perspective hurts. Live in the present, that's what I say, one step at a time, and forget nostalgia, forget history, forget the sketchy chain of loss, attrition and disappointment that got you into bed last night and out of it this morning" (Boyle 111). The passage reflects Ty's immersion in the present that allows him to block out the inconvenient truth—which would mean acknowledging without cynicism that his activism was futile and that dramatic environmental changes have taken place. It also allegorically refers to and critiques the wider context of contemporary society's preoccupation with the now. Indicating that a perspective of the past would painfully relate to us what has changed, Boyle's novel exposes that the current obsession with the present is grounded in convenience and ignorance. However, as the novel continues, it becomes obvious that such shortsightedness is not allowed for long. Even though Ty would like to forget the past, it keeps interfering with his present, forcefully reminding him that a temporal disconnect is only an illusion. Memories as well as people from his past thus begin to play an important role in Ty's present, urging him to take into consideration a longer temporal perspective.

That the novel counters a culture of forgetting is most prominently demonstrated in the story within the story—i.e., the book that journalist and activist April Wind is writing about environmental activist (and Ty's daughter) Sierra Tierwater, who died at a very young age. In the beginning, Ty is not convinced of the purpose of such a book for "posterity" (16), since this posterity is a world "turned to shit" with "people who know no more about animals—or nature, or the world that used to be—than their computer screens want them to know" (7). However, once he holds the text in his hands, he appreciates "the crisp sound of paper, the printout, the stuff of knowledge as it used to be before you could plug it in" (261). Sierra's biography not only materially and symbolically reminds him that the past is not detached from the present, but it also becomes a way of challenging the overwhelming cultural amnesia of a society where people are considered historians if they can remember what happened twenty years earlier (Boyle 217). In addition to the importance that is given to the remembrance of the past, the book about Sierra also opens up a way of thinking about the future, of reflecting on what kind of a posterity one can hope for and what kind of stories would be helpful to be passed on.

Sierra's story takes on such an importance that even the narrative pattern of the novel is interrupted. As past events reach their climax, they are also reclaiming their space in the present, interfering with the otherwise so clear-cut narrative that distinguishes between past and future by chapter divisions. Thus in Ty's first-person narration, otherwise strictly focused on the telling of the present (so on the events happening in 2025), memories of Sierra's tree protest intrude, and the reader learns the truth about what has been foreshadowed throughout the entire novel (Boyle 261–4). After spending three years in a tree, Sierra, who defied all efforts of discouragement by the logging company, accidentally slips and falls while talking on the phone to her father. As Ty finally no longer denies this tragic event of the past, which is now materially manifested in the biography, the "concrete thing, undeniable, a weight in [his] hands" (261), it invariably becomes part of Ty's present and the readers' future. Holding the book in his hand allows him to recount past events in his story that is otherwise so focused on the present. Subjective and very personal memories here come to signal their subversiveness toward a culture of forgetting. Elaborating on the differences between history and memory, Christine Boyer has insightfully pointed out, "Operating only in fragments, memory is an art that connects disparate events; it is formed on the tactics of surprise, ruptures, and overturnings that reveal its true power" (136). The moment of surprise is emphasized by the structural collapse of time distanciation in this particular passage from the novel. Ty's personal memories thus function as a way to reclaim a part of history (Sierra's protest), which may otherwise have been forgotten because it was unsuccessful. Furthermore, connecting past, fictional present and readers' future, this passage exposes and underscores that future environmental destruction is a result of past (in)action. Sierra's environmentalist defeat and death in the lush forest is then immediately juxtaposed to Ty's world of colorless forests where wilt and decay predominate the scenery. At the same time, Ty's personal loss is accompanied by the loss of biodiversity, thereby also foregrounding the intricate connection between human time and environmental time.

While environmentalist rhetoric often draws upon the past as part of an elegiac mourning for what "great nature" has been lost, this notion is complicated in *A Friend of the Earth*.[5] Though Ty reverts to lamenting that there is nothing he really wants "except the world the way it was" (Boyle 260) with his daughter, parents, and first wife still alive but also with the "doomed and extinguished wildlife ... put back in their places" (ibid.), the accounts of his past reveal that none of the environmental strategies succeeded in preventing this destruction from happening. Instead, ironically, the very nature that the characters are trying to save becomes the cause of their deaths (Kerridge, "Narratives of Resignation" 89). Thus Mac is killed by the lions he is trying to save, Sierra falls from the tree she wanted to prevent from being felled, and Teo—"environmental superman" (Mayer, "American Environmentalism" 228), lover of Ty's third wife, Andrea, and protector of nature in general—is struck by a meteor. Richard Kerridge thus concludes, "the novel looks back not at a cause narrowly defeated but at one that was always going to be lost" (88). Since not even environmentalists can free themselves from either the capitalist culture influencing them or from selfish personal motives, the environmental cause that they are pursuing is also weakened.[6] In the end, looking at the devastation in 2026, even Ty has to admit that after years of eco-tage as a "human hyena"—a nickname that was given to him after he almost poisoned the fresh water reservoir of Santa Barbara (217–218)—he has accomplished nothing, "absolutely nothing" (270). The novel ultimately demonstrates then that environmental protest will not lead to achievements if it is not heard and supported by the rest of society. Irony is used in the depiction of the environmentalists' failures in order to underscore that while a different situation should have been the expected outcome, this change would have required a communal effort. It also calls attention to the efforts of the past that were futile and reminds readers of what has not been achieved. Irony is thus purposefully employed here to break with the deception by shifting baselines and unforgivingly reconstructs the temporal perspective.

Though dystopian and harshly critical of humans' role in the destruction of the biosphere, Boyle's work refuses to fall into lamenting a lost static and romanticized nature, not least because the book is satirical and ironic. Even in a world where rising greenhouse gasses have caused global warming (185), destruction is omnipresent, and life on earth may no longer be what it used to be, the characters are not experiencing an end of nature but a world where nature still prevails. As Ty points out toward the end of the narrative, "the woods ... are coming back, the shoots of the new trees rising up out of the graveyard of the old" (274). As the shoots signal, there is hope that grief over the losses of life—human as well as non-human—can be overcome and that new forms of life may emerge. The importance of Ty's remembrance of the past is thus not necessarily about mourning the past, but functions as a rhetorical device of present disruption to bring to the fore the temporal continuity which connects past failures, present moments of crisis, and catastrophic events in the future.

Making Timescapes Visible: Narrating the Future-Present

Through the contraction (using analepsis and prolepsis) and partial collapse of time, by structurally blurring past and future, *A Friend of the Earth* permits a timescape perspective. Drawing on Barbara Adam's analysis of environmental time, Richard Kerridge elaborates that the timescape perspective puts "the timespans of ordinary life, onto which we map our personal hopes and plans, ... alongside drastically longer and shorter distances" ("Ecothrillers" 243). Especially in the case of global warming, where change does not register immediately and thus often eludes human perception, this timescape perspective goes beyond current societal "presentification" and the focus on the daily or weekly. In Boyle's novel, zooming in on the lifetime of one individual alongside the devastating environmental effects of global warming allows the reader to see and trace the emotional and personal contours of climate change. Furthermore, it shows that even in as short as time as the human lifespan, significant climatic changes are already taking place.

Even though Boyle clearly demarcates the future not only through the description of setting and characters, but also through temporal markers such as chapter headings, he also employs several narrative techniques to show the reader that the narrator's present is the reader's future. First, the timeframe chosen, 2025, is not a too-distant future from the reader of a novel published in 2000. More importantly, however, this temporal distance is interrupted on several occasions when Ty directly addresses the reader. Breaching out of the future narrative framework, he explains: "And just like you—if you *live* in the Western world, and I have to assume you do, or how else would you be reading this?—I *caused* approximately two hundred fifty times the damage to the environment of this tattered, bleeding planet" (43, emphasis added). With this metalepsis Boyle allows his narrator to transgress the fictional world, and though he thereby interrupts the fictional illusion, this is a helpful rhetorical device to convey that the fictional future is the extrapolation of the non-fictional present. As Eva Horn has furthermore argued, an important characteristic of future scenarios is that they require someone who supposedly "knows" the future and shares the information about the risks that are to be expected. Analyzing the movies *Terminator* or *Twelve Monkeys*, she concludes that they employ a narrative trick, which permits the protagonist to return from the future to the present to warn people about the imminent crisis (Horn, "Der Anfang vom Ende" 8). In Boyle's work the protagonist does not return to the present of his own fictional setting but to the present of the reader, nevertheless, functioning in that moment as the only knowledge bearer about the future, whom needs to be trusted in order to change the future.

Not only collapsing the division between fictional world and the non-fictional world of the reader, but also by juxtaposing present and past tense in this passage, Boyle moreover stresses that the reader's present is simultaneously already a moment of the past. The storytelling character of the passage, which draws readers into their future but the narrator's present, thus highlights the transformation of the current reading moment into past. This not only allows for a critical reflection on what kind of a future could have been avoided, but offers a moment of contemplation about the present. The "structurally unique 'method' for apprehending the present as history" (*Archaeologies* 288) is described by Frederic Jameson as an important feature of science fiction. As he argues, science fiction provides the critical distance and the necessary defamiliarization to the present moment—otherwise "unavailable to us for contemplation in its own right" (ibid.). Accordingly, the problem for Jameson is not only that we are immersed in the present, but more importantly that this immersion forecloses any critical reflection on the now.[7] In her introduction to the *Ecozo@* issue on "Eco-Futures," Heise, however, points out that recent science fiction complicates this key characteristic presented in Jameson's definition because some of the most recent science fiction no longer solely portrays the present as past, but rather "envision[s] the present as a future that has caught up with us, for better or for worse" (4). Indeed, most of the climate change fiction discussed in this book takes place in the near-future and suggests that many of the climatic changes believed to occur in future are already happening today or at least have their roots in the present. The present is thus not past but also already future, thereby stressing that we are already part of future environmental crises.

Boyle's novel cannot easily be classified as either science fiction in Jameson's terms, presenting the present as past, or in terms of Heise's future as present.[8] Instead, I would argue that *A Friend of the Earth* navigates between the two positions but more importantly, does so in order to provide a timescape perspective, stressing that the reader's present is intricately intertwined with the past *as well as* the future. It would thus be limiting to consider the novel's imagined future as solely a reflection on the present, especially because such an approach fails to acknowledge the continuity of the extent to which humans have been, are, and will continue to impact and accelerate planetary timescales. Furthermore, the novel aims to break precisely with Jameson's idea that the present is "*all* we have" (*Archaeologies* 287, emphasis added) and stresses that the climate change crisis requires us, after all, to think *beyond* the present. This means including the past in the retelling of the future-present relationship. Navigating between the two poles, the future and the past, highlights that the present moment is intricately connected to events of the past and the future and thus stands in continuity—something that is often disregarded by present-based, short-term thinking.

Furthermore, *A Friend of the Earth* points to the dangers of remaining embedded in the present and short-term industrial time because it impedes an awareness of slow and longer-termed environmental processes. The novel thereby demonstrates that future environmental crises are the direct result of a conflict between the dynamics of environmental time and industrial time (Rose White 4). As indicated in the introduction to this chapter, industrial time is dominating Western capitalist societies to the extent of having externalized environmental temporalities. Natural processes, "re/productive and re/generative capacities of nature" (Adam, *Timescapes* 12), do not play a role once time has become commodified and only serves to measure output. At several points in Boyle's novel, the protagonist criticizes these tendencies with his comments about capitalism's emphasis on growth. He admits that once he had also been a victim of this capitalist logic, accumulating more and more things. He also had "spent money, spoiled his daughter and watched her accumulate things in turn" (Boyle 43). The narrative of accumulation, however, is challenged by the simultaneous personal loss: the death of his first wife (who died because of an allergic reaction to a bee sting), the death of his daughter, and the loneliness that Ty experiences. Though economic time may exploit environmental temporalities (overproduction and exploitation of resources), it cannot stop the underlying natural processes. Therefore, death is still inevitable as symbolically stressed by the two deaths "caused" by nature (bees and trees). Moreover, the passage about Ty's past adherence to consumption practices is immediately juxtaposed with a description of the deprivation that Ty experiences in the future climatically changed world. The devastating weather patterns are thus postulated as the long-term environmental consequences of short-termed capitalist logic.

In fact, the future presented in *A Friend of the Earth* seems to suggest that efforts to override environmental time have been so successful that temporal variation, as humans are used to it today, no longer exists. This, however, as the novel illustrates, has not been a desirable transformation. In 2025, the term "season" can only be used to distinguish between extremely dry or extremely wet periods during which ever-present "black sky" (2) makes it difficult to distinguish between day and nighttime. These irregular weather patterns have made farming impossible and many species are extinct. Thus with meat or other fresh produce no longer available, the human diet has changed dramatically. The people surviving under these harsh conditions have to cope not only with material deprivation, illnesses caused by the lack of sunlight, and rapidly spreading diseases, but also with the emotional hardships of living in solitude, frequently leading to depression.

As birthrates decline and society ages rapidly, there are even attempts to interfere with the "natural" temporal processes of the human body: organs can be re-grown, signs of age are hidden by plastic surgery, and a distinction between nuances of young (young-young, young, young-old, etc.) has made exact age obsolete. Yet, instead of considering this development as a form of medical progress, Ty observes, "We could live another twenty-five or fifty years even. The thought depresses me. What is going to be left by then?" (Boyle 260). The control over the temporality of a lifetime then is worth nothing if ultimately this only means the worsening of the current moment of environmental crisis. In this context, posterity becomes an empty signifier, a word without meaning in a society that aims to render environmental time obsolete. Ty ironically notes that under these living conditions rats are the only species that is thriving, "multiplying like there's no tomorrow (but of course there is, as everybody alive now knows all too well and ruefully, and tomorrow is coming for the rats too)" (Boyle 6). Tomorrow, the future, is thus not something to be looked forward to, but figures as a space that is as upsetting as the present.

The possibility of returning positive value to the term "posterity" is most prominently negated by the death of Sierra—the daughter who dies before her father. While the prolongation of one's own life may be possible, Boyle's novel reveals that this ironically is not an achievement, but a punishment. By living longer, one also has to endure the consequences of one's own destructive behavior for longer. As Sierra's death further seems to stress, the intergenerational contract has been broken in a different way than expected: it is not future generations who will suffer the consequences of our current behavior; instead, environmental degradation will be endured by the generation that is causing it. This also confirms Heise's claim, explained above, that contemporary science fiction often portrays future losses as already-present losses. In Boyle's novel, climate change does not affect some distant future generations, but supposedly contemporary readers.

Finally, *A Friend of the Earth* challenges the Newtonian linear understanding of time—which underlies industrial time and which assumes that reversibility and restoration is always possible. By showing that the human focus on industrial time, through which environmental temporal patterns become externalized and nature is quantified, has significantly altered the planet's bio-physical setup, the novel also demonstrates that we can never be really independent of the underlying natural processes that enable our existence on the planet. Thus nature's invisible rhythms, though disturbed, nevertheless continue to exist. In the end of the novel, this is most prominently brought to the readers' attention, when the protagonist describes how increasing "natural" disasters have put industrial time to a halt, paradoxically leaving people with "nothing but time on their hands" (274). This, however, is a different time. Being bound to natural processes, Ty explains how "time takes hold of us and we find ourselves drifting through the days in a pattern as pure and uncomplicated as anything" (ibid.). Environmental temporal patterns thus occupy a central place toward the end of the narration and counter the idea of the end of nature. Instead they indicate that posterity may just mean engaging in a perspective that incorporates the slow and invisible underlying temporalities of the environment.

Boyle's novel explores the dangers of solely focusing on short-term industrial time by tracing the life of one individual and depicting the ways in which it is dramatically affected and constituted by his changing environment. The narrative frame thereby allows for a longer-term perspective, which is key for the reader in order to identify otherwise elusive, invisible, and long-term environmental hazards. While remaining within the timeframe of the human lifespan, Boyle's novel not only depicts a climatically changed future, but emphasizes that climate change has to be seen in its historical continuity and thus also looks to the past in its narration. It thereby becomes obvious that future catastrophes are the consequence of past inaction. *A Friend of the Earth* shows that in order to understand climate change, it is imperative to comprehend the temporal dimension of this phenomenon. It thus calls attention to the hidden patterns and cycles of nature that form the basis to our current lifestyle and that anthropogenic global warming, obscured by society's focus on industrial time, threatens to disrupt. Instead of advocating that humans attempt to master environmental temporal patterns, Boyle's work proposes living more in accordance with them and trying to create a posterity that is worth living for.

In rethinking and retelling the temporal dimension of climate change, *A Friend of the Earth* demonstrates that one needs to include the perspective of the past as well as the future. Reconnecting past, present, and future establishes historical continuity, challenges shifting baselines, and facilitates a critical reflection on the current moment of climate crisis. Finally, focusing on a reading of time in climate change fiction—more precisely, on how environmental timescales are intricately intertwined with human timescales—helps illustrate the ways in which literature participates in climate change communication. The following analysis of Jean McNeil's *The Ice Lovers* will further explore how this particular novel epitomizes the conflictual relationship between environmental time and industrial time by focusing on the specific qualities of polar time.

The Experience of Slow Polar Time in Jean McNeil's *The Ice Lovers*

Between January and April 2002, scientists witnessed the rapid disintegration of almost the entire Larsen B ice shelf (3250 square-kilometers) in the Antarctic Peninsula. This event, unprecedented in scale, powerfully demonstrated the effects of global warming in the particularly sensitive pole regions and directed the public's attention to these distant and seemingly inert places. The poles, and especially Antarctica, for a long time figured as "ends of the world"—distant, dangerous, and uninhabited. However, they have now undergone not only a physical but also a symbolic transformation. After its discovery, Antarctica symbolized a place that had to be conquered for imperial reasons, but more importantly, explorations to the South Pole came to epitomize humanity's urge to dominate nature. Staged as a battle between man and ice, many of the first explorers went to extremes on their quest to reach the South Pole and only few survived.[9] Antarctica thus for a long time predominantly figured as a vast and empty space of ice, deadly and absent of human life.

Ironically, while these glacial spaces appeared to remain unaffected and untouched, humans were increasingly having an impact on the polar regions because of rising CO_2 emissions.[10] Not only did the materiality of the ice in Antarctica begin to change, but global warming has also impacted the very idea of the place in recent years. Today, it is no longer the locus of conquest or of white wilderness, but instead "the ice is fragile, melting, ever shifting—in need of rescue" (Glasberg 222). The Antarctic has become a site that needs to be preserved at all costs because it functions as a constitutive part of the climatic balance on earth. However, this affective change toward the ice—once dreaded, ferocious opponent and now lamented, beautiful landscape—is also based on the general understanding that humankind has indeed conquered this last bit of nature. It is no longer dangerous for humans, as they can comfortably travel to Antarctica to see the last glaciers breaking apart while mourning the loss of a world of ice.[11] With a mixture of excitement and sadness, "polar tourists" watch and hear the ice crushing into the underlying sea.

Apart from powerfully demonstrating the changed "nature" of Antarctica, the well-mediatized breakup of the Larsen B ice shelf even more importantly served as an iconic moment that rendered climate change visible. As modern technology allows for a shortening of time in relation to space—epitomized in David Harvey's "time-space compression"—and media reports keep people constantly updated on events happening around the world, the focus lies on the present and on immediacy. Rob Nixon tellingly points out that in these times of short-term thinking, bringing the long-term into view requires an engagement with representational questions such as "how can we convert into image and narrative the disasters that are slow moving and long in the making, disasters that are anonymous and that star nobody, disasters that are attritional and of indifferent interest to the sensation-driven technologies of our image-world?" (3). Global warming is this type of slow disaster, registering as rising temperature graphs in scientific scenarios rather than in spectacular televised events. However, when the Larsen B ice shelf collapsed within a very short timeframe, it served to redefine the speed of climate change. It thus "recast 'glacial'—once a dead metaphor for 'slow'—as a rousing, iconic image of unacceptably fast loss" (Nixon 13). For a moment, the polar environment was then perceived in terms of time.

Yet, as explained above, environmental discourse has predominantly referred to the Antarctic in terms of place. Since the Antarctic lacked the temporal continuity of human history, focusing on the particularity of the place and the conflictual nature-human relations it proposed seemed more valuable. Also in climate change discourse Antarctica is mostly related to in terms of space—as a way to emphasize that even the most distant places are being affected, or to comment on the geopolitical conflicts that have arisen over this particular place. Similarly, some of the most well-known climate change fictions, Ian McEwan's *Solar* or Ilija Trojanow's *Eistau*, depict voyeuristic tourist trips to the Arctic and Antarctic alike and thereby rely on a rhetoric of mourning a place that will soon be lost. These works of fiction, however, fail to engage with the temporal dimension of this "loss of nature." Yet understanding the "slow violence" of today's environmental crises such as climate change requires a narrating of the dynamics of this *change* and thus a re-grounding in time.[12]

Jean McNeil's novel *The Ice Lovers* explores the temporal dimension of climate change in the supposedly static setting of the Antarctic. Though the novel's protagonists are icebound and thus confined by their seemingly unchangeable environment, the story focuses on aspects of change and natural transformation during and after this confinement. The author thereby provides insight on the slow changes of global warming and renders the *experience* of environmental time visible. The novel illustrates that affect plays a crucial role in noticing the shifting registers of critical changes and explores the emotional dimension of crossing climatic thresholds. Finally, *The Ice Lovers* not only asserts that short human timescales cannot be thought of apart from environmental temporalities such as the long timescales of ice, but, similar to Boyle's novel, reconstructs the relation of past, present, and future, emphasizing the importance of imagining not endings—but other futures.

Closing the Time Lag

The narration in Jean McNeil's novel alternates between the years 2011 and 2016 and zooms in on the lives of two female protagonists (Nara and Helen), who separately come to overwinter in Antarctica. *The Ice Lovers* thereby provides a female perspective on the experience in the Antarctic, a discourse so far dominated by male accounts. Nara, a marine biologist, comes to Antarctica more out of chance than by intention. Replacing another researcher who has unexpectedly dropped out, she commits to spending a year in the Antarctic in order to examine the temperature threshold at which certain marine animals in the Antarctic will become extinct. During her stay on base, Nara falls in love with Alexander, a glaciologist whose models demonstrate that the ice in the Antarctic is no longer forming like it used to. However, rejected by Alexander and deprived of other social relations on base—since her only friend, the pilot Luke, has left—Nara is facing a long and dark winter. The more Nara feels isolated from the people on base, the more she seeks out a connection to the planet in the belief that the ice is talking to her.

Four years later, Helen, a journalist and historian, travels to Antarctica in order to find out what has led to Nara's death during the so-called "ice-incident." However, when the international community declares a state of emergency due to the effects of worsening climatic conditions, she, too is forced to stay and overwinter in the Antarctic. As focalization constantly shifts in the novel—from autodiegetic and heterodiegetic in Helen's story to homodiegetic and heterodiegetic in Nara' story—and as narrated time switches back and forth throughout the chapters, the lines between past and present, between Nara's and Helen's lives, begin to blur and their stories become inseparably intertwined. In several chapters, the reader is thus at first not able to make out the particular temporal context of the narrative.

Furthermore, setting the story in 2011 and 2016, the novel deliberatively challenges the notion of climate change as something taking place in the far-away future. Early on during her stay on base, Nara reads in the newspaper about the growing geopolitical conflict, "dubbed the Ice War" (McNeil 90), which was ignited by the discovery of oil resources in the Arctic after the ice had retreated significantly. While she is safe in Antarctica, she realizes that "*out there*" (ibid.), i.e., in other parts of the world, people are dangerously affected by heat waves in the summer and influenza outbreaks during the winter. Only a year later, the year in which

Nara dies, circumstances dramatically begin to change, as warmer global temperatures facilitate the spread of deadly viruses. Helen later remembers:

> Good ice year, bad ice year—this is what Antarctic veterans call them. A good ice year is when the ice forms to the thickness of three or four feet. A good ice year only comes one year in ten, now. *The sea-ice incident happened in a bad ice year, also the plague year, the summer thousands died from the virus.* (33, emphasis added)

Alluding to the higher frequency of bad ice years, the passage emphasizes the accelerating rate of global warming. It furthermore illustrates that the world "out there," as Nara calls it, is also heavily dependent on the natural cycles shaping Antarctica.

However, Antarctica is not as safe as Nara believes it to be. When Alexander falls off a cliff during one of his expeditions, he finds himself surrounded by water and is no longer able to walk back to base over the ice. As environmental cycles are interrupted and the ice is no longer forming properly, people on base are also increasingly endangered. The novel thus explores climate change as a complex environmental phenomenon that is taking place in the Antarctic as much as in other parts of the world, stressing that "time and space are linked after all" (213), as Nara points out.

The severity of the disruption of environmental time cycles is further exemplified in human activities and daily routines, which, usually taken for granted, now come to a halt. Looking back at the year 2012, Helen recounts how warmer temperatures have facilitated the spread of an infectious disease, which had a profound impact on London's population. Helen remembers, "The first sign [of the 'erratic currents'] was the rubbish. It accumulated, fermented and stank" (McNeil 176). With an increasing infection rate, workers no longer manage to remove the trash, which is thus left in sight. For the first time, people are confronted with the decomposable waste they produce and with the fact that many of their interactions with the environment are not easily reversible. Furthermore, this passage in *The Ice Lovers* enables a timescape perspective by closing the "time lag between polluting action and visible symptoms" (Adam, "Re-vision" 97). The situation worsens when it becomes known that the virus can be transmitted through the air, and, as a consequence, people are required to stay inside. As the invisible underlying environmental rhythms—the weather patterns that enable and structure human routines—are thrown out of balance, humans are reminded of their own limitations and dependencies on these dynamics. Confined to her house, Helen is not even able to open the window let alone to interact with other people and thus has no choice but to watch "the plants on [her] balcony wither, yellow, droop, then die" (McNeil 176). The lockdown forces her into an observer position, in which she cannot do anything but wait and watch the virus continue to spread and ultimately kill her husband. Confronted with the promptness of death, Helen notes how the "virus has a brain, an engine, a black heart" (177) and thus works at its own pace and temporal patterns. The virus, finally, challenges human self-determinism and contemporary society's belief in humankind's dominance over nature and control of environmental time. In this context, Stacy Alaimo has insightfully pointed to the ethical and political possibilities that are opened up by "imagining human corporeality as trans-corporeality, in which the human is always intermeshed with the more-than-human world [because it] underlines the extent to which the substance of the human is ultimately inseparable from 'the environment'" (2). In addition to making environmental temporalities visible, Jean McNeil's book also serves as a literary example for Alaimo's transcorporeality by highlighting the "material interconnections" and inseparability of natural processes and the human body.

Furthermore, the invisible virus epitomizes how the slow violence of climate change can be turned into sudden violence. The virus's rapid destructive force stands in stark contrast to the presumed slow incremental dynamics of climate change and effectively illustrates to what extent the long timescales of geophysical processes have been disrupted: As these processes are accelerated, human timescales are also significantly altered. In fact, the pandemic demonstrates once more that human time and environmental time cannot be thought of separately but are intricately intertwined. Alexander's explanation of the dynamics of thermal melting serves to further illustrate this: "'Ice melts from within, but also from the outside. When these two processes happen together it's the end of the line for the ice. That's what's so interesting,' he said, '*its life cycle is just like ours, birth and decay, although over inhuman timescales*'" (212, emphasis added). While a continuous build-up of CO_2 emissions over time has led to a rise in global temperatures and thus threatens the long-term life cycle of ice, the human life cycle is also shortened by the rapid spread of the virus. Furthermore, as Helens explains, "the backwards logic of this particular sickness meant that if you were young and healthy you were more likely to die … because the virus overwhelmed the young body more quickly, spurred on by the vitality of youth" (177). Oblivious to the natural temporal processes of birth, growth, and decay, the virus affects those who are healthy and young most severely, thereby making the interruption of natural cycles even more dramatic.

The 2012 outbreak of the virus is not a singular event, but as the reader learns from Helen's recounts two years later, plagues have become the norm due to further serious disruptions in natural life cycles. Luke, the novel's key observer of the natural transformations in Antarctica, also confirms that there have been significant changes over the past ten years. Having worked as a pilot in the Antarctic for over a decade, his survival has depended on his paying close attention to the processes of nature in this particular place. He observes,

> In the ten years he'd been flying, he had watched the ice retreat, stealthily at first, then with a consuming dissolution. All around them, the glaciers were pulling back, folding into themselves. Ten years before, the glacier above base still ran down to within a few feet of the runway and the average daytime temperature in summer was minus two. They had to look out for dew-points, then, for icing in low cloud. These days the problem was fog. (80)

While these transformations may not have been obvious to the one-time visitor, they are clearly visible to the man whose profession relies on closely watching the changing natural conditions. Paradoxically, Luke, an airplane pilot who is not physically confined to the Antarctic ice and instead flies far above it, serves as the close observer of the transformations of nature. The temporal dimension is shown in the passage above in a before-and-after manner in order to emphasize the dramatic changes that have taken place. Luke's observations foreshadow that if ten years of Antarctic climate changes have been significant enough to lead to deadly diseases in other parts of the world, the situation is likely to worsen over an even longer timescale.

The novel effectively collapses short human time and long environmental time by illustrating how even the long time spans of ice are intricately intertwined with our daily routines. It explores human beings' interdependence with natural processes and illustrates how "we are locked into nature's silent pulse … our activity and rest alternations, cyclical exchanges and transformations, seasonal and circadian sensitivities are tied to the rhythms of this earth and its solar system" (Adam, "Re-Vision" 92). Without having to engage millennial timespans, the author provides an innovative narrative format to render the long "dying" of ice visible by re-conceptualizing it in terms of the rapidly spreading virus. The novel thereby also sheds light on the experience of the "slow violence" of climate change.

Icebound—Living Polar Time

A great part of *The Ice Lovers* engages in a timescape perspective that gives way to the temporal features of living in ice. It zooms in on Helen and Nara's "overwintering" experience and thereby further highlights human's embeddedness in the slow temporal patterns of nature. Helen's comment tellingly reveals, "In facing an Antarctic winter she had been committed to time in a way that was new for her" (309). Being confined to a place that forces humans to adhere to natural cycles and patterns, Nara and Helen thus find themselves confronted with a re-temporalized nature. They have no choice but to live by the seasons. As Nara observes, "in the Antarctic, winter is a door which, once closed, cannot be prised open" (166), stressing that nature's temporal patterns determine whether or not the people on base are able to leave the place. This, however, does not mean that time comes to a halt, but rather that transformations during the winter in the Antarctic are much slower and barely noticeable until the turn of the season, when winter changes to summer with abruptness. The novel thus again challenges assumptions that industrial time can dominate all natural processes by epitomizing, through its focus on life at the South Pole, how human life always depends on environmental time.

Exploring the experience of time, the novel presents time as something more than just a quantifiable resource. It displays the physical and psychological effects of time on the human life cycle. The differences between summer and winter could not be more extreme than in the Antarctic. While in the summer, "'the days passed fleeting and swift, as in cities'" (198), as Helen reads in the diary of one Australian explorer, in the winter "the day is night" and "the days last forever" (ibid.). Elaborating on the joy experienced in the summer, the narrator recounts,

> Time is circular. The sun goes round and round in the sky ... Yesterday and tomorrow are as one, artificial delegations, unruled by daylight. Nara has no shadow for weeks and finds that she does not need it, this ghostly chaperone which has accompanied her all her life one way or another, in all other places. (67)

Constant daylight leads to plenty of outdoor activities and a jolly atmosphere on base. With the day-night time cycle set on hold, there are barely any "natural" resting times and people frantically attempt to structure their days around rituals which include celebrations, skiing events, research, and tourist activities. In this "oxygen-rich" environment, it seems that anything is possible and one feels unburdened and relieved of one's memories of the past, and even content with solitude—Nara, for instance, is not even bothered by the loss of her most intimate companion: her own shadow. However, Nara also senses that this is a false optimism since "*we are not meant to live in a continuous present tense*" (75). Alluding to society's obliviousness to the devastating long-term consequences of climate change, this warning calls attention to the dangers of living too much in the present. What the people on base ignore is "a darker current" which runs "beneath this voluntary amnesia" (75) and which would reveal to them that the ice is "dying in the summer, each year a little more" (76). Not unlike their contemporaries in the rest of the world, the people on base live the illusion of dominating the processes of nature and being able to enforce human patterns of time management on nature. However, when Nara is surprised by bad weather during a field trip, it becomes obvious that even in the Antarctic summer, environmental time imposes restrictions. During what pilots call "tent days," Nara and Luke are forced to hold out in their plane because of a snowstorm. Stranded on ice, they are "waiting for rescue, for the weather to shift, waiting to be told what to do—days when time became an obstacle" (54). This scene vividly illustrates how the two are exposed to the variability of nature. Thus incarcerated, all they can do is wait for time to pass.

Yet, it is the winter that most prominently reminds the people on base of the limitations that natural cycles pose. Overwintering in the Antarctic illustrates how human biological clocks are deeply entrenched in the wider context of nature's activity as well as rest cycles. While humans may attempt to live according to artificially constructed routines and aim to break natural temporal patterns, the human body still adheres to seasonal dynamics. Accordingly, Helen describes the severe effects that the deprivation of light during the Antarctic winter has on her body: "I can't look at photographs of growing things ... my body says it is May and this should mean a lightening of body and spirit, a growth of hair and nails ... but instead my body is shutting down. My skin is dry, my hair limp. I feel no yearnings of any kind, not even hunger" (199). Due to less daylight and thus a reduction in melatonin release, Helen's circadian rhythm is significantly interrupted and as a consequence, her sleep-wake cycle, her production of hormones, and her digestive functions are disturbed. Once again, the novel here illustrates how bodily functions heavily depend on external cues and environmental rhythms.[13]

In addition to these physiological changes, the seasonal incarceration also causes psychological deterioration, especially for Nara. In the winter, "She is an open wound of loneliness" (236), increasingly depressed and psychologically unstable. She not only begins to hear voices, but also increasingly feels detached from her own body with the consequence that she refuses to eat because her stomach does not feel like a part of her any longer (236). Finally, Nara has a nervous breakdown and spends several days in the clinic on base. When she wakes up again, she feels better, but still believes she has the special ability to communicate with the Antarctic. The experience of slow time in the Antarctic winter has ultimately left her with a different sense of self and a heightened awareness of her profound dependency on her environment.

Even though wintertime in the Antarctic may often be perceived as an obstacle, it is not an unchangeable or static period of the year. Stressing that change is always a fundamental aspect of environmental time, the novel thus emphasizes the notion of *temporary* confinement. The use of terminology by the officers on the ship, who when trapped in ice are *ice-bound* but "never *stuck*" (McNeil 135), highlights this notion of slow but constant change. The officers acknowledge that the situation is temporary and that it is just a matter of time until they will be able to move on again. The novel thus portrays nature as always being in flux. Therefore, rhythmicity, tempo, and changeability of nature are given prominence in the narration. Already the opening lines of the novel, quoted here at length, emphasize a dynamic idea of nature, key to an understanding of the environmental time of climate change:

> The ice shelf looms in the windscreen. Behind [Luke], *sea-smoke is rolling* in. He loves its ermine swirl, even if it impedes his visibility … Underneath the plane the ice in the Weddell Sea is finally *breaking up*; from the air it looks like muslin—the thin, lacy frazil ice, *ice flowers sprouting on the perimeter of meltwater*. It is an absolutely dingle day in December 2016. This is his favourite moment in the *Antarctic calendar*, early in the brief summer, watching the *ice loosen its grip* on the continent. The sun is out so the contrast is good, he can see the sastrugi and the *fractures drawn* on the surface by crevasses. (McNeil 11, emphasis added)

Flying over the Antarctic, Luke describes the dynamic nature of the scenery he encounters. This short passage highlights Antarctica's context-dependent environmental rhythms and patterns, emphasizing the beauty of the transience of ice as well as the power and agency of nature in this particular moment. The reader's attention is hereby drawn to the importance of seasonal change and the opportunities that this holds.

The possibilities as well as restrictions imposed by the seasonality of nature function as more than just a temporal framework determining human actions. They also allow for a deeper understanding of environmental time, described by Barbara Adam as "time sensed but rarely talked about, a time that is internal to phenomena and constructed in interaction, a time associated with *natura naturans*" (*Timescapes* 32). Invoking the Spinozean idea of an actively transforming nature versus a static and readily modeled nature (*natura naturata*), Adam aims to redirect our attention and sensibility toward the generative and interactive processes within nature, which often remain removed from our senses. *The Ice Lovers* similarly alludes to an immanent temporality of nature and tries to re-establish the connection between nature, time, and sensibility. Various moments throughout the novel, comparable to the one discussed above, thus provide an active and transformative image of nature:

> It is winter now. All around the continent the ice tightens its grip … The Antarctic icefield is a self-reinforcing system: land, water, air—they all depend on the ice. Ice makes more of itself, becomes more itself. In becoming, its being is strengthened. The pack forms. Water is slow: the Antarctic circumpolar waters circulate the world oceans every one thousand years, creating velocities of cold, eddies, thermal columns which govern the great currents of the oceans—the Gulf Stream, the Humboldt—and which affect the planet's climate. In this process, the sea ice is key … Less is known about the continent, its processes, its intents, than anywhere else in the world, apart from the deep oceans. (172–3)

McNeil brings the slow natural transformations, which are invisible to the human eye but which serve as the underlying current of our very existence, to the fore. More importantly for the narrative structure, she turns nature itself into one of the protagonists of the novel. The polar ice is thus more than just a setting and functions as a key character for the narrative, interacting with the other characters in the novel and, notably, contributing to their emotional constitution.

Feeling Time

The Ice Lovers explores how in order to notice nature's shifting registers, one needs to reconnect to what has become detached from sense and sensibility. When Helen opens Nara's overwintering files on the computer, she discovers that Nara's distress and nervous breakdown had been caused by her belief that the "planet had a will, that it was adjusting its settings" (McNeil 23). Nara not only heard voices, but she also had visions of ghostly figures—spirits of the planet—warning her that the "pace of the world was quickening" (230), and that the earth was having difficulties in trying to keep up with this acceleration. She heard the planet's pulse going faster and beating louder, reminding her that the planet was heating up and the ice was transforming quickly. Interestingly, neither Nara nor Helen questions the existence of the planet's voice. Nara, the scientist, is bewildered yet somehow believing: "She knows these things are real … but yet she does not believe in them. At the same time, she believes in them, but she cannot know them; there is no experiment she can devise to confirm their existence" (230). She finally acknowledges that even though some things may be beyond scientific explanation, they are nevertheless real. Nara begins to trust her own sensibility in registering the great planetary transformations taking place. Because of her deeper connection to the environment, she feels that something is wrong one morning when "the universe had taken a darkling cast" (272), which prompts her to go outside in the search of Alexander. Indeed, it is this premonition that ultimately saves Alexander's life.

The novel illustrates that the Antarctic not only motivates the protagonists to listen and to closely watch nature's transformation, but also to feel it. Consequently, affect and emotion play a crucial role in recognizing the vitality of natural processes. Melissa Gregg and Greg Seigworth describe affect as "those forces—visceral forces beneath, alongside, or generally *other than* conscious knowing, vital forces insisting beyond emotion—that can serve to drive us toward movement, toward thought and extension, that can likewise suspend us (as if in neutral) across a barely registering accretion of force-relations" (Gregg and Seigworth 1). By giving agency to natural processes, McNeil alludes to those "intensities" that arise in the spaces in-between humans' encounter with Antarctica. This becomes most prominent in David's description of his ship passage to the Antarctic. David, who is called the "polar man" in international politics and becomes Helen's husband at the end of the novel, describes how "he could feel the continent long before it arrived" (118). At a later point he elaborates, "What he feels for the place is unmistakable; that is what love feels like, to him, always" (121). David's comment reveals here that affect forms a constitutive aspect of his experience of the Antarctic. Reading this passage in light of Alaimo's concept of transcorporeality as well as affect theory highlights the invisible but nevertheless material exchanges between bodies and their environments. Feeling then allows nature to matter.[14]

As the story continues, the intricate emotional and physical connection between characters and the ice is made even more explicit, especially because of the novel's emphasis on "thresholds." It is not only Nara's research objects, the sea animals, that are increasingly distressed due to higher temperatures, but every human also "ha[s] a different threshold at which they will stop functioning" (87). Nara, too, is physically and psychologically exhausted after she has been rejected by Alexander and is left behind by Luke, who flies off and abandons Nara for four hours in the vast whiteness of Berkner Island during the "ice-incident." When Luke returns to rescue her, their plane crashes and Nara is too weak to survive. As Helen later writes, "*The species which will die out are those which are unable to adapt in time*" (22). Nara was not able to adapt in time. Just like the ice, she literally disappears since her body is never found. Though reminding the reader of emotional and climatic thresholds alike, the book leaves room for hope by presenting ways to imagine an alternative future in which these thresholds can be avoided.

Re-writing the Future after the "Crystal Embrace"

The Ice Lovers not only draws parallels between Helen and Nara's overwintering experience on base, thereby bridging the temporal disconnect between past and present, but it extrapolates their stories into the future. Working as a journalist, Helen describes that she used to write stories about endings and thus about war and devastation. She points out that she believed that "We need these narratives of termination in order to confirm our worst suspicions about ourselves, the fleetingness of our lives, the power of destruction we have, the harm we are capable of wreaking on all and anyone who strays too near" (23). However, after coming to the Antarctic and doing research on Nara's experience on base, she discovers that the climate change story cannot be told as an apocalyptic scenario. Instead, the story about Nara and ultimately about environmental degradation is quiet, subtle, slow, and very complex. Moreover, it is not necessarily a story about the end, but about *krisis*. This becomes obvious when David muses:

> Change was not a simple manoeuvre, performed by temperature rises, flux, ablation, water from ice and ice to water. In human lives, change required crisis which in Greek, as he had never forgot … meant *decision*. The *krisis* is the moment when you walk down one path and forsake the other. (118, sic)

By placing changes in nature in immediate relation to human action, this passage, instead of ascribing a notion of ending to crisis, invokes the idea of chance and the possibility for alternatives. It also demonstrates that the moment of *krisis* is characterized by uncertainty because one cannot know in advance if the chosen path is the right one.

However, in the beginning of the novel this uncertainty poses a great problem to many of its characters who are obsessed with wanting to know the future. Prior to her departure to the Antarctic, Helen, who is "sick of uncertainty" about her life, seeks out a fortuneteller. Yet the woman tells her that while she can see some things about the future, because "they have already formed in the same way that crystals congeal" (18), others things flow in the river of life around them and they "could go either way" (18). Similar to Helen's desire to know what lies ahead, Nara is preoccupied with the future. As "part of a vast puzzle of researchers sent to discover clues to the future" (23), she believes in the scientific ability to know the future. While she conducts experiments concerning the temperature threshold for marine animals' extinction, she is also fascinated by Alexander's ice modeling. When she finds out that the models indicate that the ice is no longer forming, she begins to photograph snowflakes before they melt. However, this newborn interest does not derive from a desire to preserve what will soon be lost, but rather comes from her belief in the planet's own will. By looking closely at the flakes, she hopes to see a truth, to recognize "a grand design." Helen comments that after all Nara, too, "was trying to see through the crystal, into the future" (246). Finally, David's brother, a trained economist is also portrayed as someone who is trying to foretell the future. David ridicules the enormous scientific efforts of designing future scenarios as well as the banks and governments who come to visit his brother and "his crystal ball." Again relying on the idea of *krisis*, David explains that it does not help to have "all the futures, floating in front of us" (297) because the problem still remains one of choice—of deciding on the right kind of a future without knowing the future. The crystal of the fortuneteller, the crystals of the snowflakes, and finally the scientific scenarios all represent ways to map the future, to provide insights on "what will it be like, to live in a world without ice" (315). However, as the novel juxtaposes various methods of prediction, none is presented as the one that provides any relief or certainty. The future remains open and, as Helen learns over the course of the story, decisions in life will have to be taken without knowing what the future will bring. Yet, this is not necessarily negative. She realizes, "I no longer want to know the future. I've learned that the future is far more slippery than we think. And there is another danger: once we know the future, we feel it owes us to turn out as we have been told it will" (331).

Arguing against knowing the future for the purpose of controlling it, Helen rather embraces *krisis* as the crucial moment in the present, which allows for a contemplation of various possible futures. This is illustrated in her way of dealing with her research project. Since the Antarctic Sea, where Nara supposedly drowned, never releases her body, Helen never finds out what had really happened—if Nara had been concussed after her plane crashed and drowned or if she had deliberately walked "away over the ice not for rescue, but to embrace death" (306). Yet Helen realizes that while "[she] may not be able to excavate the fleeting moment and the truth it holds locked in its crystal embrace" (332), she may still be able to tell Nara's story. She begins to write a novel about Nara and thereby reveals a future "had the crystal river flowed in another direction" (331) and Nara had survived. Helen thus turns from being a journalist writing about the present to being a historian writing about the past to being a novelist writing an alternative past and future. Proposing a different future scenario, Helen's book about Nara highlights that while the future cannot be predicted, it can be imagined. This imagined future "is only marginally less real than the one that actually happened" (332), because it is equally plausible. More importantly, Helen's book reveals that even small changes in the past and present can provoke very different future outcomes. McNeil thereby stresses that, while various futures are possible, they are all bound to the present moment of *krisis*, to thinking about and then deciding on a path for the future.

Thus, the future that Helen imagines for Nara is not disconnected from the past and present. On the contrary, *The Ice Lovers* illustrates that there is no such a thing as a separate temporal timeframe. Instead, past, present, and future are always intricately intertwined. The proleptic aspects in the novel, the fictional flashforwards in which moments of present and future become conjoined, therefore function as part of a longer timescape perspective, for which the past is equally important. The changes in Helen's professional career—turning from a recording of the past to a rewriting of the future upon her return from Antarctica—further stress that "To know the future you have to look at the past" (211). Finally, the novel's focus on the experience of time in the Antarctic, a place in which "time never ends, nothing is ever past" (115), negates a temporal disconnect and instead magnifies the temporal overlap of existences. In addition, Helen and Nara's stories are not only connected because of their similar experience on base. As Helen looks into Nara's past, there are moments in which Nara's spirit seems to leap out into Helen's present: Nara's appearances in Helen's mirror, as well as Nara's writing in the overwintering file, evoke her presence. Furthermore, the narrative structure of the novel effectively collapses the time distanciation between the two women's stories. Often it is not clear to the reader in which time frame the next chapter or even the next paragraph takes place—whether this is Helen's or Nara's story.

This "overlapping of existences" is further illustrated through references to David's great-grandfather, who had barely survived the Antarctic polar exhibition with Ernest Shackleton. While David comfortably sails past Elephant Island, he muses that "what separated him from a starving man, exposed, frozen, lost on that narrow shore, was only time" (122). However, as much as time separates them, the two men nevertheless remain connected. Obsessed with the Antarctic and going back to it again and again, David's destiny is determined by the experience of his great-grandfather—and David, too, over the course of the narration will be trapped on ice. The novel's intertextual references to Shakleton's diary further parallel the three different, yet similar, experiences of being icebound and stress a temporal continuity. The *mise en abyme*-like storytelling then exemplifies the intricate connection of life stories across time and fosters an awareness that no present or future generation will live disconnected from its past.

As this close reading analysis of the novel's temporal dimensions has aimed to show, even the long timescales of ice are intricately intertwined with short human timescales. Consequently, as environmental temporal patterns, which so far have sustained human life, come to be disrupted due to climate change, human time is also severely affected. Furthermore, by emphasizing the transient nature of a supposedly static place such as the Antarctic, the novel magnifies the agency of nature and the power of environmental cycles.

T.C. Boyle's *A Friend of the Earth* as well as Jean McNeil's *The Ice Lovers* both set their climate change stories in the near future and thus within the temporal horizon of their contemporary readers. In addition to countering the assumption that climate change is something that will take place in the far-off future, the novels emphasize that it is important to become attuned to a different temporal dimension: environmental time. In trying to understand the climate crisis, turning to the slow and invisible environmental temporal processes on which human time is based is crucial. *A Friend of the Earth* and *The Ice Lovers* bring these processes to the fore and thereby provide a timescape perspective. Furthermore, they remind readers how the past, present, and future always stand in continuity. Tellingly, the narrative structures of the novels discussed here aim to bridge temporal disconnects in order to enable thinking beyond the present, a reflection on the past as well as the future.[15] This connection serves to emphasize humankind's role in shaping the futures to come. While both authors provide a dystopian vision of future, their works nevertheless leave room for hope. In the current moment of *krisis*, another path may thus still be taken. However, as the discussion in the following chapter shows, seizing that moment of *krisis* and taking action is complicated in a world of manufactured uncertainty.

Notes

1. In his insightful essay on messianic time, Giorgio Agamben points out that the representation of time is often spatially bound (graph, line, etc.), which makes time representable but not necessarily thinkable. He finally turns to poetry to overcome the antinomy between the representation and thinking of time and to get at what Guillaume has called "operative time"—"time in the act of constructing itself" (Agamben 3). I would argue that the climate change fiction discussed here engages in this representational challenge and aims to get a grasp of this elusive category of "temporality."
2. Adam further explains, "Industrial time … is centrally structured to (a) the invariable beat of the clock, (b) the economic commodification of time and (c) the scientific use of time as measure of abstract motion" (*Timescapes* 11).
3. Fish management is the prime example put forward in this context. As Saenz-Arroyo has elaborated by looking at three generations of fishermen, the reference frame for fish depletion in each generation was set at the starting point of their careers and did not include an

awareness of earlier fish stock extinction (Saenz-Arroyo and et al.). While the concept has been criticized in marine biology studies because of its somewhat naïve assumption that it is possible to return to a formerly "pristine" state of nature (Hobday 393),which, however, proves very difficult to be determined, it has served as a helpful entry point of discussion for studies related to practices of remembrance in social sciences and the humanities.

4. With the term "dictatorship of the now," Shellnhuber is referring to the exploitation of past and future alike (for example, of fossil fuels) in order to satisfy our exuberant demands of the present.
5. In her essay on cultures of extinction, Heise indicates that environmentalist discourse has so far heavily relied on "declensionist narratives" and their call for an awareness of the loss of "beautiful" nature ("Cultures of Extinction" 50). See also my discussion Chap. "Reimagining Time in Climate Change Fiction".
6. In contrast to Kerridge, however, who analyzes Boyle's novel as a "narrative of resignation" that takes environmentalists' failure for granted, I would argue that it is precisely the book's irony which undercuts the otherwise dystopian setting.
7. Coming back to Shellnhuber's dictum of the "dictatorship of the now," Jameson's idea is helpful, because it provides insight as to how the dominance of the present can be disrupted. Furthermore, Jameson's elaboration accounts for the power of fiction to provide a necessary distanciation to the present moment and to emphasize historical continuity.
8. Moreover, as discussed in Chap. "Scaling Climate Change: The Transformation of Place in Climate Change Fiction", in climate change fictions genre conventions are blurred. It is thus difficult to define works such as Boyle's *A Friend of the Earth* as science fiction at all.
9. Two different expeditions to the South Pole, one led by Roald Amundsen (Norway) and one by Robert Falcon Scott (England), were the first ones to be successful in reaching the pole in 1911 and 1912 (The National Science Foundation). However, on his return trip, Scott and his team were surprised by bad weather and died. Also Ernest Shakleton's famous trans-Antarctic expedition almost met a deadly end when their ship got trapped in the ice and they were icebound for several days (see Mills).
10. As elaborated in the British Antarctic survey (2009), "Stronger westerly winds in the northern Antarctic Peninsula, driven principally by human-induced climate change, were responsible for the marked

regional summer warming that led to the well-publicised retreat and collapse of the northern Larsen Ice Shelf." The survey furthermore states, "The temperature in the Antarctic Peninsula has risen by almost 3°C in the past 50 years causing some of the smaller ice shelves to melt."

11. It is helpful in this context to turn to Catriona Mortimer-Sandilands's discussion of melancholy natures. Instead of an active engagement with the loss of nature due to anthropogenic destruction, Sandilands argues that trips to the Antarctic or other endangered places can be considered as a form of nature nostalgia, which incorporates "environmental destruction into the ongoing workings of commodity capitalism" (Mortimer-Sandilands 333). In a society that is not really able to value those non-human environments that have been lost, grief is projected onto certain commodified sights or objects (national parks, the Antarctic, etc.).
12. Rob Nixon closes in on the often invisible and disregarded violence caused by current environmental issues such as climate change or toxicity and sheds light on threats often overlooked in our fast-paced lives. He specifies that by "slow violence" he means "a violence that occurs gradually and out of sight, a violence of delayed destruction that is dispersed across time and space, an attritional violence that is typically not viewed as violence at all" (2).
13. In *A Friend of the Earth* medical progress is postulated as a way to overcome precisely these environmental dependencies. Yet, unsuccessfully as Ty's repeated complaints about weather caused ailments suggest.
14. I am alluding here to Alaimo's word play in *Bodily Natures* (2).
15. The climate change fictions discussed here then also seem to counter Beck's claim that the determining power of past becomes obsolete in a risk society, because of an overpowering dominance of the future (*Risk Society* 44). Boyle and McNeill's novel both point to the dangers of such disconnect.

Manufactured Uncertainty: Climate Risks in an Age of "Heightened Security"

Mitchell Zukor is afraid. In fact, he has always been afraid—of a supervolcano eruption burying the United States under a layer of ash, a nuclear showdown with China, a new form of the black plague, or another terrorist attack. He thus spends his days with "logic games" (Rich 3), calculating the probabilities of all kinds of worst-case scenarios, which seemingly provides him with some sense of security. During one of his web searches on future disasters, he comes across an ominous New York–based company called FutureWorld, a consulting firm that purports to predict future disasters so that its clients will know what the future is going to cost them. Mitchell is instantly intrigued and soon after contacting the manager, he begins to work there as the company's first consultant—as a futurist. However, unlike his superior, who engages in worst-case scenarios purely for financial gain, Mitchell's fear is real. Profiting from the fact that clients see the urgency in Mitchell's eyes "of great danger coming on" (Rich 31), FutureWorld quickly experiences an upsurge in business. While both Mitchell and his clients become increasingly absorbed in the mathematics of disaster, an actual worst-case scenario occurs.

In his latest novel *Odds Against Tomorrow,* Nathaniel Rich not only critically challenges the growing importance of probabilistic calculations for political and especially economic decision-making, but also raises questions of how to navigate through and adequately respond to varying uncertainties in everyday life. The author depicts how the protagonist—

A. Mehnert, *Climate Change Fictions*,
DOI 10.1007/978-3-319-40337-3_5

Mitchell Zukor, who is obsessed with future modeling and is trying to avoid catastrophe all his life—is one of the few people who anticipates the flooding of New York City, but is still not prepared when it happens. While Mitchell regularly guides his clients through "a special feature on terrorism" (Rich 72), scenarios of public health scares, and natural disaster sequences, he does not give any importance to more immediate threats. These include the extreme period of drought that his friend Elsa, who lives in the countryside, describes in her letters to him, and the long-lasting heat wave that he also experiences in the city. Only when the drought spell is finally broken by heavy precipitation does Mitchell realize that the weather anomalies he had predicted in his scenarios were true. Yet, knowing that after such a long period of drought the land is unable to absorb the kind of heavy rain that Hurricane Tammy, which has taken a steady course toward New York City, will inevitably bring with it, his boss and his co-worker urge him not evacuate. So he stays and exposes himself to severe flooding. His feared worst-case scenarios, then, are no longer part of his nightmares but have become reality.

Rich's novel ultimately depicts a society confronted with threats beyond its control: a "risk society." Coined by German sociologist Ulrich Beck, the risk society denotes a society that finds itself endangered by the self-inflicted hazards of its own processes of modernization (Beck, *Risk Society* 14). Setting dynamics into motion whose consequences lie way beyond calculability and control, contemporary society is faced with ecological, financial, biomedical, and a multitude of other risks. Most of these risks do not adhere to nation-state boundaries and pose a global threat (Beck, *World at Risk* 10,15).[1] In one of his more recent publications, Beck insightfully distinguishes between threat, risks, and manufactured uncertainty, while acknowledging that these terms at times overlap ("World Risk Society" 293).[2] Contrary to threats, risks—while also describing the anticipation of catastrophe—presuppose human decision-making, certain calculability, and thus responsibility assessment (ibid.). Looking at the etymology of the term, Beck points to the origins of the word "risk" in relation to the emergence of the first insurance companies in maritime trading.[3] The monetary notion that adheres to risk is thereby prominently brought to the fore. Beck even goes so far as to state that in the seventeenth and eighteenth centuries, probabilism and risk assessment began to replace moral reasoning, since actions were becoming framed in terms of the likelihood of their occurrence (Beck, "World Risk Society" 296). This measuring of probabilities of future events in order to prevent economic risks plays into what political scientists Michael Dillon and Luis Lobo Guerrero have termed the "commodification of contingency" ("Biopolitics of Security" 281). This profit-making element of risk assessment will also play an important role in the discussion of the Rich's work in this chapter.

In addition to this managerial and monetary concept of risk, Beck postulates that contemporary discussions often revolve around "manufactured uncertainties." These particular risks are "incalculable, uncontrollable and in the final analysis no longer (privately) insurable" ("World Risk Society" 293). Beck elaborates that manufactured uncertainties ultimately challenge existing ideas of certainty and rationality in contemporary Western societies, especially because they break with assumptions about their controllability (qtd. Yates 99). They are less manageable, work on a global scale, can be either latent or fast-moving, and are part of a complex network of risks. Decision-making under the conditions of manufactured uncertainty is thus particularly difficult because one does not choose between safety or risk but between various risky alternatives, none of them easily measurable (Beck, "World Risk Society" 298).

Yet, this also poses questions about the representation of risks, because, after all, risks are always anticipated and thus require some form of narrativization, visualization, or mediatization. How then are uncertainties produced discursively and made visible, let alone made experienceable? In the context of climate change, graphic visualizations such as the so-called emission scenarios have so far been the dominant means to project future threats and to make climate risks visible. Combining biophysical models with assumed social and political trends, these scenarios are commonly defined as "plausible, challenging and relevant stories about how the future might unfold" (Pulver and VanDeever 1). As such they provide important data on the possible greenhouse gas trends for the future.

As climatologists, however, very succinctly point out, climate scenarios do not function as a prediction of the future because nobody can know how humans will act and react since "that is a matter of choice" (Archer and Rahmstorf 126). Even though there is a very overt notion about uncertainty and probability in these statements, the scientific scenarios work under a problematic context. As products of the scientific community, they are understood and expected by the general public to present "factual truths," i.e., certainty about the future. Media studies scholar Birgit Schneider pointedly explains this controversial status. She describes that while these scenarios aim at an intervention and at providing plausible insights on what the future could look like, they are caught within a system of assumptions ascribed to their discipline and scientific workings. What could be called "science fiction" is discussed in terms of its truthfulness and thus contested by the general public, even though the importance lies in taking these fictional realities seriously (86).

The misconception about the functioning of scenarios has made them particularly vulnerable to critique. It is not surprising then that discussions on climate change have revolved around the accuracy of science and the assertion that climate science is settled. While the IPCC states in its 2007 *Synthesis Report* that "the warming of the climate system is unequivocal" (30), it cannot say anything about its causes or impacts with absolute certainty.[4] In order to make any knowledge claims, the IPCC then uses a variety of epistemic modifiers such as "very likely" (>90 %), "unlikely" (<33 %), or "high or medium confidence," but its method of presenting the research results clearly illustrates that science is not a discipline that aims at "being settled."[5] Returning to Beck's elaborations on contemporary risk society, uncertainty, however, should not serve as an excuse for inaction. On the contrary, especially since risks can no longer be compensated for in times of manufactured uncertainty, precautionary action may be the only preventative measure (Ewald 273). This would mean to embrace uncertainty and doubt and to nevertheless take action to avoid the possible worst.[6]

However, for a long time the public debate on global warming in the United States has focused on the accuracy of data instead of the meaning and consequences of these findings on policy measures, such as cutting CO_2 emissions. For climate change skeptics, this focus on uncertainty has helped to promote their political agenda, declaring that as long as the science is not settled—i.e., presented as facts with a 100 percent certainty—there is no reason to reduce emissions of anthropogenic greenhouse gases. Therefore, powerful anti–climate change lobbies managed to create a public image of a scientific controversy where one does not truly exist (Boykoff and Boykoff 125–6). This deconstructive approach toward science, emphasizing its uncertainty, has cast doubt on the results of climate science and paved the way for a denial of the problematic materiality of anthropogenic climate change. Even Bruno Latour wonders if he had been wrong in the past by trying to demonstrate the "lack of scientific certainty" ("Why Has Critique Run Out of Steam?" 227), upon realizing how this "constructivist technique" can be easily misused. This cynical misuse of constructivist critique by climate change skeptics is exemplified in the following section from a 2003 piece in *The New York Times*:

> Most scientists believe that [global] warming is caused largely by man-made pollutants that require strict regulation. Mr. Luntz [a Republican strategist] seems to acknowledge as much when he says that "the scientific debate is closing against us." His advice, however, is to emphasize that the evidence is not complete. "Should the public come to believe that the scientific issues are settled," he writes, "their views about global warming will change accordingly. Therefore, you need to continue to make the lack of scientific certainty a primary issue." (qtd. in "Why Has Critique Run Out of Steam" 226, sic)[7]

In light of this approach, it is no wonder that the discussions around climate risks have thus become a testing ground for critical science studies and their engagement with science. More importantly, the controversy about climate scenarios illustrates how power relations and the question of who declares risks play a decisive role.

To date, the complexity of climate change and its great variety of possible effects have seemingly only increased societal uncertainty about it. In addition to problems of scientific ambiguity, abstract simulations on a global scale make it difficult to illustrate the problem in a way that can be translated into everyday life. It is no wonder then that laypeople find themselves constantly doubting their own actions and concerns toward global warming. Also, communication scholars such as Susan Moser have pointed out that it is more difficult to communicate with the public about climate change, with its complexity and uncertainty, than it is to discuss other environmental problems. She explains that it is extremely challenging for most people to grasp, let alone process, the complexity of climate science, and even if they do, mitigation strategies raise similar questions of uncertainty (Moser 35). Under these circumstances, apathy and paralysis are the most likely reactions.

Different narratives and visualizations may thus be needed to shed light on how to navigate a world of manufactured uncertainties. In trying to answer how threats, risks, and uncertainties may be more effectively narrated, Beck points to the importance of literature. He acknowledges that "literature concentrates the uncertainty of the future into narrations and thereby makes it experienceable and perhaps even biographically and politically understandable" (Beck, "World Risk Society" 298). Also, other scholars such as Ursula Heise and Susan Mizruchi have pointed out that literary fiction plays an important role in putting information about invisible risks into an intelligible and socially meaningful order. This socially meaningful order allows not only for the selection of what kinds of risk need to be contemplated, but also points to possible causes and consequences, and defines actors and victims, thereby making the invisible visible and disclosing a path through this complex web of uncertainties. Furthermore, Mizruchi explains, it is in particular the ethical and affective dimension of risks that make them "especially suited for fictional representation" (119).

One prominent example of literary fiction about climate change risks is Michael Crichton's best-selling thriller *State of Fear*. He illustrates how uncertainty is manipulated and deliberately employed in order to meet certain political goals. Yet, in his book climate change is ultimately reduced to a sham made up by a group of eco-activists. Crichton uses several graphs and climate models in his novel and arranges them strategically to foster doubts about the relation between increasing global temperatures and increasing greenhouse gas emissions. This distortion of facts is in itself not remarkable for a work of fiction. However, the controversy about the book deserves some attention. While the thriller actually calls for a more straightforward division between science and politics, its aftermath evoked the opposite response. In 2005 Republican Senator James Inhofe called Crichton before the Senate Committee on Environment and Public Works in order to testify on the role of science in environmental policy making. This caused an outcry in the scientific community and ignited a debate about who has the authority to discuss climate change. More importantly for the analysis in this chapter, however, is the issue that Crichton raised during the hearing by asking, "What should policymakers do with research that is unverifiable?" (Crichton). While this question also proved to be the starting point for his novel, it refers back to the general discussion about how to navigate through a complex system of uncertainty.

The Crichton example reveals that the process of making uncertainties visible—in scenarios or narratives—always involves a certain set of beliefs and forms part of larger power structures. In *State of Fear*, eco-activists ignore research that does not support their cause and not only deliberatively manipulate data but also engage in violent action to prove that climate change is happening. To this end, they systematically approach and "misinform" influential public figures and con them out of millions of dollars. While Crichton seemingly engages critically with the issue of uncertainty in climate research, his book—ultimately relying on genre characteristics and limitations—falls back on a very obvious good versus bad rhetoric and thereby undermines its own question about the verifiability of research. As the eco-activists are turned into ruthless villains of the story, there is no longer any doubt about the validity of their claims.

So, the plot of the thriller and even the author's attitude after its publication show that Crichton positioned himself along the lines of climate skeptics by discrediting the scientific research done so far, calling for more research and stalling action. Crichton's work resonated and epitomized US public controversy about global warming at the beginning of the millennium.

Contrary to Crichton's work, which employs uncertainty as a means to justify political passivity, other works of recent climate change fiction also critically engage with issues of uncertainty but take climate change seriously, nevertheless. Nathaniel Rich's *Odds Against Tomorrow*, introduced at the beginning of this chapter, juxtaposes climate change risks with the predominant need in contemporary society for predictability and security. Moreover, the novel provides insight into the power structures of risks, on who decides if risks are defined as such and thus profit from them, thereby revealing that risks are not necessarily "democratic," as Beck claims (*Risk Society* 36).[8] Finally, Rich depicts how risks are experienced and dealt with on a personal level, offering ways for characters and readers alike to navigate through a world at risk. For the protagonist in *Odds Against Tomorrow* only an unmediated relationship to the land, and no general formula of risk probabilities, can provide safety at last.

Communicating Uncertainties and Risks

Rich's novel contrasts the personal experience of risks with large-scale risks that are projected in various future scenarios. It thereby introduces an element of manufactured uncertainty into the discussion that has often been neglected so far—that is, the affective dimension of fear. The protagonist of the novel, Mitchell Zukor, is seemingly afraid of everything. Especially in the first chapters, which describe Mitchell's condition more in depth, the reader learns that his anxiety even provokes physiological reactions—fainting, heavy sweating, and temporary paralysis. Svendsen's analysis in *A Philosophy of Fear* can help frame the discussion about this aspect of Rich's novel. Svendsen argues that fear is an emotion determined by social, cultural, and physiological processes. More importantly, he adds, "The core of fear is the assumption of a negative future situation" (39). This explains Mitchell's obsession with wanting to predict the future in order to know if the future will be as bad as he fears it to be. When he is a university student, Mitchell is already chronically afraid and becomes a regular patient at the university's health services because of his constant suspicions of potential health problems. While Mitchell's fellow students are convinced that he is psychologically ill, his mental state seems to vary between paranoia and a heightened awareness of risks. The novel hereby interestingly poses the question of how much risk awareness is sane—an issue that will be discussed more in detail at the end of this chapter.

Ultimately, the protagonist's chronic anxiety can be distinguished from paranoia because it does not happen in a state of delusion. Instead Mitchell "rationally" approaches his fears and attempts to "contain" them in risk models. As he explains to his mother at one point in the novel, "Everyone has fears. It's just a matter of controlling them" (Rich 76). By calculating probabilities of worst-case scenarios, Mitchell then not only gains control over his fears, but also evaluates which risks to take seriously and which ones to discard.[9] This is illustrated in a comic manner when the protagonist overcomes his fear of taking the elevator. Working on the seventy-fifth floor of the Empire State Building, Mitchell soon recognizes that while taking the steps up to his office is a good exercise, it will also cost him a fresh suit every time. After calculating the chances of dying in an elevator accident, he realizes that they are "one in 10.44 (repeating) million—about equivalent to the odds of dying from a dog bite, according to the National Safety Council odds-of-death chart he kept in his wallet" (46–7). As a consequence, he starts taking the elevator, but crosses the street whenever he sees a dog. However, while calculating these worst-case scenarios may provide Mitchell with a sense of safety—as he puts it, "fearing the worst usually cures the worst" (Rich 76), they do not actually prevent disasters from happening.

The beginning of the novel illustrates that Mitchell's dread of large-scale disasters is emblematic for an entire generation, which not only grows up in fear but also has to continually face actual catastrophe. What comes to be dubbed as "Generation Seattle" witnesses a particularly destructive earthquake wiping out the entire city of Seattle. As one of Mitchell's college friends later recounts, nobody, not even Mitchell, had seen this disaster coming. The panic raised by this event marked an entire generation for whom "Both the best and the worst suddenly seemed possible" (11). Even a couple of years after the event, while wandering through the streets of New York, Mitchell notices that more and more people were feeding on and profiting from this culture of fear. In the aftermath of an actual disaster, people seemed more inclined to believe that there are still greater risks to come and thus all over the city, preachers of apocalypse find large audiences. The protagonist tellingly observes, "A feeling was building. An urban malaria, a future-affected anxiety disorder. Whatever kind of disease it was, it had become infectious" (51). Mitchell's fears were no longer singular and individual, but had quickly spread to become collective fears. As Svendsen argues, fear is contagious and, "If someone becomes afraid of something, this fear has a tendency to spread to others, who in turn spread it further. This may occur even though there was initially no rational basis for the fear" (14–15).[10]

In *Odds Against Tomorrow* future *angst* becomes omnipresent, but at the same time society adapts to a state of fear and the thought of facing threats beyond its control. It increasingly dwells in crisis. Thus, the particularly long and severe heat wave in the novel intensifies the general feeling of fear and anxiety "which had settled like a poisonous cloud over the country after Seattle" (75), but does not provoke any action or preventative measures. Not even Mitchell, who focuses in his business meetings on climatic irregularities and especially drought scenarios, draws the connection between these developments and their possible consequences. Only when the heat wave is suddenly interrupted by heavy rain and when Hurricane Tammy threatens to make landfall on the East Coast, does Mitchell, tipped off by a climatologist's comment, start to think about deluge rather than drought scenarios. This illustrates that while trying to control his fears, Mitchell had become immersed in the construction of one risk scenario after another, but had lost a grip of reality and real-life threats.

Interestingly, Rich does not once mention the term "climate change" in his novel. While the heat wave and the subsequent flood are never directly explained as phenomena of climate change, the reader is, nevertheless, made aware that of all the other risks that Mitchell fears, climate change involves the only risks that he should have taken seriously. As also Mitchell realizes toward the end of the book: "A few more years of these new meteorological patterns, a few more disasters, and every person on the street would be able to speak intelligently about drought, methane pollution, UV poisoning. The intricacies of planetary collapse would be general knowledge" (254). Due to first-hand experience, people would thus no longer need any prediction models to confirm to them that the threat of climate change was real. Crichton, in contrast, builds his story around strengthening the idea that "global warming is the *theory* that increased levels of carbon dioxide and certain other gases are causing an increase in average temperature" (96). While equally skeptical about prediction models as Rich, he denies the possibility that climate change can have devastating effects on the basis that "no one has been able to predict future climate with accuracy" (295).

Yet, Rich's novel also critically questions the assumption that future predictions can "insure" the world against disaster or provide safety. This is most prominently brought to the fore by the unusual relationship between Mitchell and Elsa Bruner. The protagonist first encounters Elsa in college, during one of his many visits to the university's health center. Upon leaving the center—as usual without a serious medical condition—he glances at Elsa's file and finds out that she has Brugada, a rare heart disorder, which may and usually does lead to sudden death. Shocked, Mitchell realizes that Elsa personifies his worst fears, "She's a walking worst-case scenario. How does she get out of bed?" (10). However, Elsa does not seem to be afraid and deliberately exposes herself to life-threatening situations. Mitchell, knowing of her condition, keeps an eye on her and ultimately saves her life. After college, the two stay in touch as pen pals even though they take very different directions in life. Working as a successful futurist in New York, Mitchell is bewildered because he does not understand why Elsa dropped out of college and chose to build and live in an eco-village far away from any medical facility. Elsa, on the other hand, wonders if Mitchell is happy and urges him in her letters to critically re-evaluate his career path and personal obsession with worst-case scenarios. Yet the key question that connects the two is never asked directly, and Mitchell only dares to pose it to himself in silence: "*What's your secret, Elsa Bruner? Why aren't you afraid? What do you know that I don't?*" (42, emphasis orig.). For Mitchell, Elsa's behavior simply speaks against any logic of survival since he believes that "evolution ruled against the fearless" (62). Elsa, however, explains that she had chosen this particular kind of lifestyle because it allowed her to follow her impulses more fully. She asserts that, "Rationality has made a mess of this world. Rationality isn't helpful or useful, and it's certainly not exciting" (64). Elsa accuses Mitchell of getting too caught up in the calculation of risk scenarios and of thereby losing touch with reality. She ridicules him because he keeps a plastic bag full of money in the freezer in order to feel safer. Finally, Mitchell's ignorant attitude and non-action in the events leading up to Hurricane Tammy confirms Elsa's suspicion, which she expresses in one of her letters as: "More fear = less curiosity about the actual world?" (61). Elsa insinuates that fear may not only be paralyzing, but also that fear externalized in risk scenarios possibly prevents one from perceiving more immediate threats. Fears, according to her, are then mostly constructed and result from uneasiness instead of curiosity toward the unknown. Therefore, risk scenarios may assign a certain probability to fears and give the illusion of controlling uncertainties, but they do not prepare a person how to behave when actual disaster occurs. Also Mitchell's mother remarks doubtfully, "These scenarios are abstractions" and encourages him, "Put yourself in the world. Be a person of action" (77). Yet, since for Mitchell these scenarios *are* the real world, *Odds Against Tomorrow* epitomizes the conflict of whether or not there is a distinction between "objective" and "subjective" risk.

By showing how risks are communicated as part of worst-case scenarios, which find particular resonance in society because they are conceived in a culture of fear, the novel ultimately demonstrates that risk perception and assessment always depend on human judgment. Rich also illustrates that a heightened awareness of risk is not only a reaction to the hazards of modern life, but also derives in part from a social dynamic of an over-preoccupation with risks. Drawing on Furedi's argument may help to clarify this dynamic. He explains, "Disasters and catastrophes have happened throughout history, but the reaction to these events has varied according to the mood that prevailed in society at the time" (6). Comparably, risks are perceived as such and have gained such prominence because they are constructed at a particular point in time, in which Western societies believe themselves to be confronted with dangers threatening their very existence.[11] Moreover, by pointing out that risks are determined by their social context, Furedi concurs with Mary Douglas and Aaron Wildavsky's elaborations on risk and culture. Highlighting the process of selection in risk discourse, Douglas and Wildavsky argue that risks are always culturally dependent and, consequently, "Each form of social life has its own typical risk portfolio"(8). Moreover, the anthropologists point out that risks are always closely related to moral judgments and to identifying victims exposed to risks as well as culprits responsible for producing them.

Rich's novel indicates that contemporary US society is particularly receptive to the risks of large-scale catastrophes. Furthermore, the media plays an important role in the construction of those fears. In the novel, Mitchell alludes to its discursive power, "It was astonishing how much bad news was generated every day. You only had to pay attention—subscribe to the right newsletters and academic journals—and you could see the information accrete, like matter spiraling around a black hole" (Rich 65). This passage reveals that the mechanisms and articulations of risks solidify the feeling of crisis in society. It also alludes to the difficulty of taking threats seriously, especially at times when society faces such a great variety of discourses of fear. Adding to the problem of deciding which risks to take seriously, Western contemporary societies no longer rely on an individual alertness to immediate dangers but rather "consume information about risks and ways to rely on agencies that reduce them" (Michael Shapiro).

Rich's novel shows that relying solely on agencies of "risk management" may be problematic because their operations are based on a commodification of risks but do not help in the coping with disaster. While Mitchell is initially absorbed in worst-case calculations because they paradoxically provide him with a sense of safety, over the course of the novel those scenarios become part of an "exhilarating cycle of doom" (79). In his job as a futurist, Mitchell has the time and resources to engage in research on every different kind of disaster. He realizes that this knowledge, when channeled into scenarios, not only makes him feel safe but also bestows him with a position of power. Continuing his disaster research, "He [now] used the numbers as a salesman would—to recruit new clients, to catalyze their fears. It had become a game to him. FutureWorld had transformed him from a neurotic paranoid into something much more powerful: a businessman" (85). His own fears but more importantly his clients' fears are thereby converted into a commodity.

The Commodification of Uncertainty

Depicting how the protagonist, working as a risk consultant for the company FutureWorld, deliberatively employs fear in order to make uncertainty matter and to thus draw financial gain out of it, Rich critically turns to the sector that profits most from a culture of fear: the insurance and financial consulting industry. As these industries operate at the intersection of risks and security and constantly have to negotiate between the two, they come to value the individual's life in monetary terms, equaling security with the possibility of generating future revenue. In his insightful analysis on *Insuring Security*, political scientist Luis Lobo-Guerrero exemplifies the close relationship between insurance practices and biopolitics in the co-production of security.[12] Drawing on various past as well as present-day examples, he argues that insurance practices are not only always immediately related to the capitalization of life, but also closely interact with sovereign forms of power. Lobo-Guerrero critically examines security and the processes of insuring security by pointing out how the individual is subjectified. He further explains this biopolitical effect, "the inscription of value into life, the classification of life as valuable or invaluable as a result of its insurability and the production of securities through the capitalization of life subjectify individuals in relation to their circulations" (Lobo-Guerrero 130). Life insurance, the monetary compensation for the loss of life, brings this most notably to the fore. Security measures in contemporary risk societies are thus not aimed at erasing uncertainty, but rather at best and most profitably managing insecurity. The strategic "manipulation" of risks and likewise security is dependent on and influenced by the institution that produces and "regulates" the information, and it involves not only defining who is at risk but also who or what poses a threat.

Rich's novel explores these dynamics of capitalizing life and critically reflects on how risk and security practices have become part of market dynamics. In *Odds Against Tomorrow*, major catastrophes such as 9/11 have led insurance companies to stop their reparatory payments, as risks seem more and more difficult to predict. Still, people looked for someone to repay them for their losses with every newly occurring disaster. This dynamic is best illustrated in the novel's description of events in the aftermath of the Seattle earthquake. Family members and business investors alike filed lawsuits against the companies that endangered their employees' lives by letting them work in a disaster-prone city. They were looking not only for someone to blame, but also for someone who would to pay for their loss. In response to their demands, courts finally ruled that businesses were to be held responsible and had to pay for the dead, thereby converting "blood ... into treasure" (Rich 15). Life at that point started to only matter in terms of its repayment value, as the narrator's comment exposes, "The loss of life, though regrettable, could be overcome. It was the loss of capital that brought the chief executives to their knees" (14). As a consequence, companies in other disaster-prone areas of the country began to worry about potential financial losses. Also the financial consulting company that Mitchell starts to work for right after college, with its office inside the Empire State Building, is concerned about their vulnerability to natural disasters. Thus, one of Mitchell's first assignments is to calculate the company's financial loss in the case of catastrophe by assessing the price of every employee's life. Upon realizing that his own value to the company is very low compared to the value of one of his colleagues, Mitchell worries if annihilation was the only thing that he could expect of his future. This reveals how he himself has internalized the idea that life only matters in terms of revenue.

While insurance firms no longer make payments to clients after large-scale disasters, the novel demonstrates that since "chaos breeds financial opportunity" (29), new business models of managing uncertainties emerge quickly.[13] FutureWorld is one of the companies that has found a niche in the risk management market and is now profiting from disaster.

While business owners are usually liable to pay for any future losses due to catastrophes, FutureWorld makes use of a legislative loophole that allows them to act as a "hired scapegoat" (Rich 26). As such, they "specialize in minimizing losses that may result from unforeseen or worst-case-imaginable scenarios ... predict all possible future outcomes, highlight the most grievous, and explain what options might be available" (20). Most importantly, they work on the basis of a legally implemented "defense to liability claims," which postulates that every property owner of a building with an occupancy of at least 200 people who has shown enough effort to make his building secure and has spent a "substantial sum on precautionary measures" (28) is indemnified. Mitchell realizes that this business strategy is ingenious because as long as businesses could prove that they had been advised on possible disasters, there was no way to hold them liable for losses. Rich's novel here exemplifies the commodification of uncertainty and illustrates how merely knowing about the possibility of hazards suffices as insurance for financial loss. This describes what according to security studies expert and political scientist Michael Dillon are the characteristics of our modern understanding of risks. He explains, "Risk makes contingency fungible and commodifies the exposure to danger and opportunity for advantage which, together, now describe what risk is" (Dillon 64). The importance thus lies in making uncertainty matter for monetary reasons. Relating back to earlier discussions of Beck's risk society, Dillon's argument gains even more relevance. In a society faced with numerous manufactured uncertainties, insurance businesses become immediate beneficiaries. Indeed, Guerrero states that "with US$ 18.5 trillion of assets under management, or 11 per cent of global financial assets, the insurance industry is world's largest economic industry" (1).

Rich picks up on this profitable business with uncertainty and shows in *Odds Against Tomorrow*, how risks are no longer feared as threats but rather become windows of opportunity. Tellingly, the logo of FutureWorld is a sketch of an open window, figuratively inviting clients to look through it in order to get a glimpse of the future. Depicting in detail how Mitchell's fear turns into fascination, the novel explores an often-unacknowledged aspect of risks: their close relation to thrill. The more information he acquires about possible disasters, the more he thrives on them. Furthermore, these doomsday scenarios also become an important part of who he believes himself to be, as illustrated in the following passage:

> The bad news brought a rush of excitement; it fortified, too. It reached an intimate part of him. It didn't merely feed his fears, it also fed his

> fascinations. The information had a way of seeping into his higher thoughts. After a while he began to feel that he was the information. He went further afield, into doomsday prophecy and eschatology. *It was tremendous fun.* (70, emphasis added)

As Mitchell comes to understand that he himself has the power to construct risk scenarios, his fear is converted into the thrill of being able to control others by frightening them. He no longer suffers from thinking of worst-case scenarios, but, for the first time in his life, he uses the fear they generate strategically as a sales tool. Revealing that any information can be—and in Mitchell's case, is—presented as potential danger, the novel critically comments on contemporary tendencies to manipulate risk discourse in order to make a profit. Uncertainties are then deliberatively conceptualized in terms of danger, and insurance measures are promoted and sold as a response to provoked fears.

Furthermore, Mitchell's own joy in creating risk scenarios plays into Svendsen's claim that risks have become part of entertainment in present-day Western societies (15)—they are exciting because they are frightening. Even though Mitchell's projections become increasingly apocalyptic, more and more clients seek out FutureWorld's service, which tellingly shows people's fascination with disaster. While clients uncomfortably squirm in their seats during consulting sessions, they are, at the same time, not only thrilled by the stories of doom that Mitchell relates to them, but are also becoming increasingly absorbed in the stories of disaster. To Mitchell, this is a normal reaction of people living in a culture of fear. He explains, "Frightened people didn't want bromides, expressions of hope, happy predictions. They craved dread, worst-case scenarios, end times. What would the future cost them? They wanted to hear that the price would be exorbitant" (109). This passage postulates that information about the future is almost immediately related to adverse outcomes—that is, people expect things to go wrong. According to Furedi, thinking of risks in negative terms is a consequence of anxieties about present problems (18). He concludes, "The very meaning of risk is shaped by how society regards its ability to manage change and deal with the future" (18). Following his argument, one can maintain that Rich's fictional society sees itself confronted with too many complex, uncontrollable problems and thus disasters seem to be inevitable.

Finally, by exposing the constructed nature of risks, and commenting on the structures that bring certain threats to the fore, the novel shows that making uncertainties matter is always closely related to the power to control people.[14] While Alec Charnoble and Jane Eppler, Mitchell's colleagues at FutureWorld, construct and use risks to manipulate people for financial gains, others such as the religious leaders that Mitchell encounters on the street of New York City employ risk scenarios in order to make their followers conform to a specific way of life. Mitchell, on the other hand, engages in risks for still other reasons. Already in the beginning of the narrative, he realizes, "If you planned for disaster and none ever occurred, you were a fantasist. But if a disaster you predicted did come true, then your life had meaning. You were a prophet" (32). Also the novel's intertextual reference to Becker's *The Denial of Death* reveals to the reader that Mitchell's fear of death is the result of the fear of not having done something in life for which one can be remembered. By creating risk scenarios, he hopes that one day one of these projections will come true, making him memorable because he has foretold the future. Risks scenarios in *Odds Against Tomorrow* are not then presented as a means to make the world safer or as a way of encouraging followers or clients to take action in the face of possible disaster. Instead, they are used as part of selfish plans for power and profit.

As clients become lured in by Mitchell's increasingly absurd worst-case scenarios, paradoxically, Mitchell does not care about *what* might actually happen but is only interested in communicating his own fear of annihilation:

> He could feel it intuitively: disaster was real, and it was coming fast, like an asteroid plummeting from the sky. Perhaps it was an asteroid plummeting from the sky. Something giant and obliterating was rapidly encroaching. What did it matter whether it were a bomb or an earthquake or a drought? His Cassandra tales grew more persuasive, more specific … Every single year, in other words, there was a ten percent chance that the species would extinguish itself. *He had no great advice to offer his clients about this fact. He just wanted them to understand the likelihood that they would be incinerated shortly.* (107, emphasis added)

While this passage not only discloses the importance of rhetoric in risk discourse, it also interestingly picks up again on the conflict between Elsa and Mitchell and thus engages the question of what makes disaster real. For Mitchell, the communication and premonition of disaster already constitutes part of its reality. The figure of the prophet further emphasizes the conflict between projection and reality, because it shows that it requires not only someone who can predict the future, but also people who believe in that prediction. Whereas Cassandra's predictions were true but nobody believed her, Mitchell's argumentation and calculations are so convincing that people indeed believe that something bad will happen. So when a catastrophe —one which for a long time Mitchell did not see coming—actually hits New York City, he comes to be considered a prophet, because his clients relate his earlier very diffuse worst-case projections to the worst-case scenario that they are now facing. Mitchell, finally in a position of power, is able to avert what he fears most: his own annihilation. However, having actually experienced the cruelty and horrors of disaster, he no longer wants to be that prophet.

Navigating through Uncertainty—Taking Risks

Odds Against Tomorrow critically exposes how the risk industry thrives in a capitalist culture, which solely focuses on maximizing profits while failing to see the real dangers and catastrophes contemporary societies are confronting. Rich thus depicts the anticipation of crisis, but also portrays how in a world of manufactured uncertainty, disaster actually occurs seriously affecting the characters' lives. The novel thereby most notably reveals how the "mathematics of disasters" potentially impede an awareness of the actual climate change crisis, and are incapable of providing advice on how to actually deal with catastrophe.

Countering the paralyzing effects of apocalyptic rhetoric, as well as the commodified excitement of gloom-and-doom, worst-case scenarios, the moment of crisis caused by abrupt change in weather emphasizes the need for action. Completely disrupting the daily lives and routines of the characters of the novel, Rich here also suggests that chaos and partial breakdown is needed in order to change habits.

Thus, after Hurricane Tammy's destructive path through New York City, Mitchell, who sat out the storm in his apartment together with his colleague Jane, understands that ultimately none of his scenarios could have prepared him for the atrocities of this disaster. He realizes, "Tammy was worse than anyone could imagine. Like all major catastrophes, it surpassed the limits of imagination ... Tammy—like Seattle—was an innovative disaster. Its horrors were unprecedented" (234). Mitchell hereby questions his own earlier work by admitting that worst cases cannot, in effect, be predicted or imagined because they are worse than anything that we have ever experienced before. Moreover, Mitchell's response to the Hurricane shows that while he knew what was about to happen and could have escaped the city, he ultimately chose to stay to protect his colleague. This foreshadows Mitchell's transformation into a caring man of action no longer paralyzed by the models of crisis.

Also the escape out of the city illustrates Mitchell's change in mindset. The water still remains very high on the morning after waves rolled through the streets of Manhattan. Even though the Coast Guard urges people to stay inside their homes, Mitchell knows that since they are running out of food they have to leave the city in order to survive. Tellingly, his only means to do so is to leave on his *psycho canoe*, a canoe made by a folklore artist who is part of a collective that is opposing a world based on rationality and, very similarly to his friend Elsa, tries to follow its impulses more (98). Paddling toward higher grounds, Mitchell and Jane confront the disastrous effects of the storm. They come across many dead bodies and manage to help other survivors, but they also witness people starting to turn violently against one another in the attempt to hoard supplies. Again, Mitchell realizes that "In his futurist calculations he had always counted on bad things happening. But he hadn't considered the brutality of it, the primitive, selfish desperation that took hold when one's life was threatened" (178). His risks scenarios—though depicting worst cases—never could have grasped the reality of life-threatening catastrophe. Compared to the actual experience of disaster, his obsession with risks and his fear becomes superfluous.

Mitchell and Jane finally make it out of the city. However, while Jane senses that this disaster works in favor of FutureWorld's business and thus wants to return to New York as soon as the city is rebuilt, the experience of actual disaster has had a profound impact on Mitchell: he is no longer a futurist and can no longer imagine what any long-term future could look like. To him, the experience of actual disaster—something that he had feared all his life—had made the calculation of worst-case scenarios obsolete. Moreover, Mitchell understands that while parts of Manhattan and Brooklyn can be reconstructed, climatic conditions will not change. The dystopian future, described in his futurist scenarios, is already part of the present. He starts to grasp the magnitude of this "present as future":

> The floods would keep coming, more and more frequently. Soon the coastal cities would lose the will to rebuild the old seawalls and levees. No one would have to pay to hear about worst-case scenarios—they'd be living them, night and day. The future would vanish as a preoccupation; the present would consume man's full energies. (Rich 195)

Upon this realization Mitchell parts ways with Jane. While he always dreamed of being a prophet, of thereby avoiding his own annihilation, he now feels the burden that comes along with "knowing the future." It is no longer about the thrill or apocalyptic excitement but involves a responsibility toward the people that are ultimately affected by crisis. Aware of this, Mitchell no longer wants to be a prophet for the horrors to come, he wants to live in the present. He thus moves to the South of Brooklyn, the Flatlands, which are not connected to the subway system and thus remain outside government rehabilitation efforts. In the middle of this "postapocalyptic wasteland" (272) he begins a new self-sustaining life, which fulfills him and rids him off his former fears. Finally following what Elsa had always suggested, "He was doing more of that, he found. Sensing. Less logic, more intuition" (273). Living off the land, and focusing on the present and everyday task of sustaining one's livelihood, seems to be Mitchell's strategy against the fear provoked by future uncertainties.

Rich epitomizes the conflict between climate risks and insured security one last time toward the end of the novel, by depicting how Jane watches from her shaded car window the flow of people that follow Mitchell's example by moving out to the Flatlands. She herself thinks, while driving toward the city where she is hiring the new "class of Cassandras," "I am *scared*" (306). While Mitchell and his followers deliberatively choose to live in a high-risk area, surely to be flooded again during the next storm, but possibly aware that there is risk-free alternative in a world transformed by climate change, also Jane's belief in insurance for security is put in doubt. There seems to be no escape from risks, but in order to not live in constant fear, as Mitchell's behavior seems to suggest, a return to the basics of life could be a solution. The close connection to the land and a sustainable and self-sufficient lifestyle "relocates" fear in the individual's body again and will thus at least reveal how to react to the risks that are to be taken seriously.

Setting most of his story in New York City and depicting how climate change conquers this supposedly invincible place, this epitome of the American Dream, by surprise, Nathaniel Rich exposes how the city, immersed in a financial bubble, has lost touch with the real world. Instead of being led by Wall Street and its constructed risk scenarios, *Odd Against Tomorrow* urges readers to critically assess what risks to take seriously. To do so, Rich suggests a redirection toward and reconnection with nature. Furthermore, the protagonist's final turn to tending the land and promoting a self-sustaining lifestyle picks up on the American ideal that a closer attachment to the land is key to environmental awareness.

I would argue that Rich, aware of the US bias against taking action on climate change, frames his stories explicitly in this recognizable American context in order to convince readers about the necessity to take action.[15] Of course, as explained in the beginning of this chapter, climate change fiction does not only represent risks but also participates in the construction of risks and uncertainties. On the one hand criticizing risk scenarios, Rich also constructs and feeds into them with his novel. However, contrary to a book such as Crichton's thriller, Rich's novel illustrates that uncertainty related to climate change does not justify its denial or serves as an excuse for inaction. So, Rich also critically engages with risk scenarios, fear mongering and specifically the commodification of risks, but to very different ends: he does not dismiss the possibility that climatic disasters may continuously occur.

Notes

1. See also Beck's elaborations in *Risk Society* for a further distinction of risks as well as his turn to the cosmopolitan dimension of risks in *World at Risk*. Social scientists have criticized Beck's concept of the "risk society" because it lacks empirical evidence. Furthermore, the concept is employed and applied to society in very general terms—Beck's statement "smog is democratic" (*Risk Society* 36) comes to mind here—and fails to acknowledge the work that has been done by environmental justice scholars showing that marginalized groups are particularly vulnerable to environmental risks. See also my elaborations on this point at the end of this section.
2. In the course of this chapter, I will employ all of the three terms mostly in accordance to Beck's distinction. However, in cases where their meanings overlap, they will be used interchangeably.
3. See also Beck's elaborations. First insurance contracts date back to Genoese trading in the fourteenth century (for a more elaborate discussion see Lübken and Mauch 9–11).
4. The parameters of certain scenarios nevertheless point to dangerous tendencies and thus suggest preferable countermeasures. In other words, the diagrams act could be interpreted as "barometers of risks"—as the measuring scale for the hazards produced in our "reflexive modernization" (Beck, Giddens, and Lash), indicating which storyline increases the risk potential. Framing climate change in this particular way, scenarios give a specific value judgment and suggest

which risk constellations of today should be avoided in order to avert a climate crisis of tomorrow.

5. Daniel Sarewitz has convincingly argued in his essay on science and environmental policy—and he is relying here on Latour—that rather than producing absolute truths and policy recommendations, scientific research is based on raising and proving hypotheses and thus producing a grand variety of scientific data. For further insight on linguistic elements and techniques used to express polyphony and (un)certainty in climate science documents, see Kjersti Flottum's analysis.
6. Frank Furedi has, however, argued that the precautionary principle has fed into the contemporary culture of fear in order to call for more security measures as well as to justify preemptive action—the war on terror provides several examples in this context (Furedi 203–206). The novel discussed in this chapter also critically depict the two sides of precautionary measures.
7. This quote has been taken from the article "Environmental Word Games," *The New York Times*, 15 Mar. 2003, p. A16.
8. I am again referring here to Beck's much disputed claim "poverty is hierarchic, smog is democratic." He continues to elaborate that with the expansion of risks "social differences and limits are relativized" and that thus risks "display an equalizing effect within their scope and among those affected by them" (*Risk Society* 36).
9. As Svendsen has elaborated, flight is a normal reaction to fear. However, this escape is not necessarily spatial and can be any kind of "barrier" put between oneself and the object (31). Mitchell's scenarios can be interpreted to function as such a barrier.
10. Examples from the past prove a case in point. Especially in the context of the US immigration, nativist reactions at the turn of the twentieth century, which led to the segregation of Asian, mainly Chinese, immigrants in San Francisco because of suspected bubonic plague, have served to illustrate how fear "spreads" irrationally.
11. Several scholars such as Svendsen, Killingworth, and Palmer have noticed a rise in publications that deal with disaster and evoke apocalypse (see also Chap. "Scaling Climate Change: The Transformation of Place in Climate Change Fiction").
12. For his discussion, Lobo-Guerrero looks at early eighteenth century England, for example, in order to illustrate how the state began to securitize its own status as the lives of its citizens were rendered

capital. He explains, "The security of citizens was premised on the capacity of individuals to translate their own lives into property, property which was the condition of possibility in the early liberal life for the exercise of political rights" (52).

13. Indeed, Rich's story picks up on a couple of issues also raised in an insightful article by Michael Lewis in *The New York Times* (2007), entitled "In Nature's Casino," in which he traces the career of John Seo, a hedge fund manager especially interested in catastrophe bonds, and thereby portrays in detail today's commodification of uncertainty. Throughout the article, Lewis describes the functioning of insurance and consulting agencies and particularly focuses on how their "culture of catastrophe" shifted after the occurrence of several major natural disasters such as Hurricane Andrew and Hurricane Katrina. He explains how new business models were found as "the big catastrophic risks of the early 21st century couldn't be diversified away" (Lewis). Insurance companies could no longer account for the high losses and the only way to handle these new catastrophic risks was by extending them to broader markets and thus onto Wall Street. This is how catastrophe bonds were created and also the reason why "big investors weren't scared off by Katrina. Just the reverse. It has led many of them to turn to Seo and others like him to make money from catastrophe" (Lewis).
14. Lobo-Guerrero also points out that there is nothing "natural" about risks but, following Foucault, risks always form part of a "rationality of governance" (2). Insurance thereby only serves as a "technology" to apply risks.
15. Taking into consideration that *Odds Against Tomorrow* was published only a couple of months after Superstorm Sandy, one wonders again about the fictional element of these stories. Also Carolyn Kormann, journalist for the *The New Yorker*, explains that novels like Rich's *Odds Against Tomorrow* can be read as social realism rather than science fiction, again bringing up the question of genre of climate change fiction.

ClimateCultures in Kim Stanley Robinson's *Science in the Capital* Trilogy

Understanding the complexity of many environmental problems today requires a research approach that transcends disciplinary divides based on the nature-culture dichotomy. Climate change, for instance, is at once a natural phenomenon, the "hot breath of civilization," as author Ian McEwan metaphorically described it (*Solar* 109), as well as an object of projection for a variety of social, political, and economic ideas. It is then the epitome for what Latour calls a "quasi object" (*We Have Never Been Modern* 64 ff.) and in his later work a "matter of concern" (*Politics of Nature* 24), since it functions as a hybrid phenomenon—simultaneously natural as well as socially and politically constructed. Countering the divide between disciplines, the term "quasi-object" implies that there is no binary distinction between nature (and the scientific construction thereof) and culture, but instead there is a simultaneous co-production of the two. Steven Connor also affirms that the invented distinction between nature and culture is increasingly disintegrating, as "more and more 'things' get drawn into social life, which will become more and more dependent upon and liable to be transformed by what it draws from and does with nature" (Connor 7). Climate change here follows in line with a number of recent phenomena (the ozone hole, mad cow disease, multi-resistant pathogens) that utterly challenge clear-cut categorizations into nature or culture.

Consequently, instead of discussing global warming solely in scientific terms, it is important to approach the topic with a perspective that builds

A. Mehnert, *Climate Change Fictions*,
DOI 10.1007/978-3-319-40337-3_6

on an understanding of humanity as being "firmly embedded within the physical climate system" (Hulme, "Cosmopolitan Climates" 38). According to geographer Mike Hulme, this requires a "cosmopolitan" thinking, an understanding of the complexity and mutual conditioning of nature and culture in a post-natural state of the world. The term "post-natural" is employed here to denote that a return to some "original nature," and thus a pre-CO_2 emissions world, is not possible. Even though the anthropogenic causes of climate change play an important role in the debate on climate, it is no longer possible to draw a line that distinguishes between natural global (non-anthropogenic) climate and artificial climate. One cannot, for example, say whether Hurricane Sandy, which hit New York in 2012, was a "human-induced" storm or a "natural" storm. Instead, a cosmopolitan approach to climate change, as Hulme proposes, may be productive. By describing climate change as "cosmopolitan," he aims to emphasize its co-produced nature.[1] Hulme, moreover, explains that he finds this term useful in order "to explore and make sense of some of the emerging sociological and psychological attributes of climate change, this '*mutating hybrid entity*'" (268). He finally concludes that the cosmopolitan—in the sense of boundary-dissolving—character of climate change challenges prominent Western binary distinctions and claims, "stability is one attribute of this hybrid climate system which will not be achieved" (270).

Climate change thus questions a reliance on former, all-too-comfortably assumed categories and requires an acknowledgement of human embeddedness in nature. While Hulme's article provides insightful arguments, it also raises questions as to how this cosmopolitan climate change can be narrated, represented, and most importantly understood outside expert discourses and theoretical elaborations. As more than half of the world's population lives in urban environments and is thus insulated from the ecological structures on which their material well-being depends, a cosmopolitan understanding of climate change seems to gain even more urgency. Indeed, while controlling the temperature of one's own apartment is understood as a matter of course, altering the global climate often lies beyond one's imagination. How then can a reciprocal relationship of nature and culture be told in a way that acknowledges climate change as a hybrid entity—as cosmopolitan, a natureculture, or what I would term a climateculture product?

Various scholars in the humanities have been challenging binary divisions between mind/body, cultural/natural and construct/material.

Donna Haraway, most prominently, introduces the term naturecultures to emphasize the inseparability of nature and culture. In his discussion of Donna Haraway's ideas, Timothy Clark summarizes the positive aspects of such an approach: "The effect of considering all the dimensions together is not to 'relativise' a 'fact' out of existence, but to demonstrate the plurality of its conditions of being accepted as a 'fact', to blur the boundaries between what is literal and what may be social/symbolic" (*The Cambridge Introduction* 160) and to nevertheless acknowledge its materiality. In addition, Emily Potter has argued, natureculture theory proposes a decentering of the human in order to flatten out hierarchies and emphasize co-constitution. A "natureculture" is then always an assemblage of a multitude of actors and interactions (Potter 3). There has been an attempt across disciplines ranging from the sciences, history, literature, etc. to do justice to this concept of naturecultures. This chapter also aims to contribute to this discussion with a specific focus on climate change. It thus analyzes how climatecultures are brought to the fore in fiction.

Turning to recent climate change fiction, which engages in a variety of challenges in representing and narrating climate change, is also insightful for the discussion of how the nature-culture divide can be overcome and what implications a climateculture perspective might have for the communication of the position of "the human" in the climate change debate. Elaborating on the idea of "literary ecology," Hubert Zapf has already pointed to the importance of literature in staging "complex dynamical processes of life on the boundary of the culture-nature interaction" ("Literary Ecology" 13). He convincingly argued that as literature transgresses disciplinary confines and the dichotomies of modernity, it produces a cultural knowledge that posits the interconnectedness of human and non-human and establishes an important basis for ethical reflection.

Kim Stanley Robinson acknowledges with his *Science in the Capital* trilogy that stories that take place in an era of climate change can no longer rely on old categorizations of nature based on a culture-nature binary. He therefore seeks to tell a different story—one that aims to render the connections between the various actors, both human and non-human, visible. This network structure makes it difficult to read nature as something apart from the human. Nature is no longer pristine untouched landscape, no longer dangerous threat that needs to be dominated. Instead, the trilogy reveals how the human and the non-human are co-shaping the planet. The following chapter highlights three aspects in the trilogy that illustrate most prominently how Robinson narrates climatecultures. First, by drawing on a complicated network of characters, Robinson stresses that climate change is an issue that reaches beyond the scientific realm and requires a multilayered and multiperspective response. He argues against a disciplinary divide and explores a way for politics and science to collaborate, making it clear that this interdisciplinary approach is crucial for action on climate change. Second, Robinson's employment of abrupt climate change serves to illustrate how humans are embedded in and an intricate part of nature—even when immersed in the urban landscape of Washington D.C.—and how the natural and the cultural are co-shaping these urban ecologies. Finally, by zooming in on the human response to abrupt climate change, the author shows that humans have not only become powerful players in times of climate change, but also that they are now confronted with adapting to constantly changing natures and with finding new ways of responsibly terraforming the planet.[2] Finally, as a writer of utopian science fiction, Robinson goes beyond declensionist environmental rhetoric and, in portraying the complex matrix of nature, provides sparks of hope in times of climate crisis.

Paradigm Shifts

While the *Science in the Capital* trilogy, similar to the other climate change fiction discussed in this book, engages with major climatically induced disasters, it conceptualizes the endangerment that climate change poses very differently. In an interview, the author Kim Stanley Robinson, a renowned figure in science fiction writing, explained that although he had been interested in the topic of global warming for some time, he had difficulty writing about it until he heard of abrupt climate change (Robinson, "Imagining Abrupt Climate Change" 5). This phenomenon, which describes the point when the otherwise slow and gradual process of climate regime is pushed over a threshold that causes sudden major changes, allowed Robinson to circumvent the problem of portraying long timescapes in the volumes of the trilogy, *Forty Signs of Rain* (*FS*), *Fifty Degrees Below* (*FD*), and *Sixty Days and Counting* (*SD*). His first book *Forty Signs of Rain* introduces a cast of characters who are scientists working at the National Science Foundation (NSF) in Washington D.C.—most notably Frank Vanderwal, who recently joined the team from California, and Anna Quibler, a statistician and mother of two children, who in the course of the novel becomes acquainted with the representatives of the embassy of the fictional island of Khembalung, which is threatened by flooding due to rising sea levels. By focusing on these characters, the story provides very detailed insights into the processes that shape scientific knowledge and exposes the difficulties scientists face in communicating climate risks to the public and to politicians. The narrative reaches its climax when a major flood suddenly hits Washington D.C., taking all of the characters by surprise and significantly disrupting their everyday routines. The sequel *Fifty Degrees Below* continues the multi-perspectival narration and describes how after this devastating flood the city's inhabitants have to endure an extremely severe winter. Furthermore, a subplot revolving around an affair between Frank and a secret operations agent, Caroline, gains in importance. The final novel *Sixty Days and Counting* depicts "a moment of history when climate change, the destruction of the natural world, and widespread human misery were combining in a toxic and combustible mix" (*SD* 5), and newly elected, eco-friendly President Phil Chase is trying his best to confront the adversaries of weather, politics, and the economy.

Robinson shows that climate change cannot be considered solely from a scientific perspective because it pervades all areas of life. By portraying in detail how the characters in the trilogy deal with global warming, the series postulates that an interdisciplinary response to climate change based on negotiation and compromise is required. Furthermore, the theme of "the network" that runs through the series underlines the importance of crossing disciplinary boundaries in the attempt to initiate a paradigm shift. In the beginning of *Forty Signs of Rain*, the reader is introduced to a Buddhist delegation from the fictive island nation of Khembalung. Due to rising sea levels, their island is threatened by inundation, and the characters Drepung, Rudra, and several others have now come to Washington D.C. to campaign for action against global warming. As Anna Quibler, a scientist at the National Science Foundation (NSF), realizes one morning on her way to work, the Khembali delegation has opened their embassy inside the same building where the NSF has its offices. As Anna thinks that "this was a strange place for an embassy" (*FS* 12), the Khembalis quickly reassure her that they have chosen this location very carefully. Indeed, the more Anna and several other NSF scientists get involved with the Khembalis, the more they become aware of the dramatic affects of climate change and begin to think about mitigation measures. Symbolically, the displaced island delegation also unsettles the lives of the characters in Washington, causing them to question their own work routines and disciplinary work ethics. At the same time, as the interests of the NSF scientists and the interests of the Khembali delegation converge, the Khembalis' presence in the NSF building spatially represents this unification.

Robinson thus signals early on in the series that climate change requires science to operate outside its institutional confines. This dynamic is further emphasized by Anna's attempts to help the Khembalis. Thinking about ways in which science could support her newly gained friends' cause, Anna finds out that the history of the NSF is characterized by a constant struggle between science and politics. While she is convinced that the NSF is doing valuable work by advancing basic scientific research, she simultaneously wonders if "they would have to consider doing something more" (FS 124). For too long, the organization had solely followed a research agenda that depended heavily on external funding and politicians' favor. Anna realizes that what was really required was "interdisciplinary research, relevant to problems of our society" (123), and for that to happen, NSF scientists had to do more than just basic scientific studies and actually set their own research agendas. Anna's colleague Frank Vanderwal, who recently joined the team from San Diego, is also very disappointed with his term as a program director at NSF, because he believes that the current environmental crisis needs more than this "*ridiculously passive position*" (*FS* 210) of solely reading and judging scientific proposals. Yet, while he condescendingly describes people passing him by at the airport during one of his work trips as "experts at filtering their information to hear only what made it seem sensible to behave as they behaved" (*FS* 209), he himself, though being aware of the problems, hides behind a facade of cynicism and brusqueness. Even though he drafts a letter to Diane Chang, the head of the NSF, in which he criticizes the agency for its minimal impact on politics, he does not engage in any straightforward action and gives in to passivity by not wanting to renew his term—at least in the first book of the trilogy. Robinson comments here on the disciplinary restraints of science and postulates that in order to have an influence on politics or everyday life, a rethinking of these disciplinary confines is required.

Moreover, zooming in on Frank's internal conflict serves to illustrate the difficulties of moving outside the comfort zone of one's own disciplinary framework. He, like Anna, a "Puritan of science" (*FS* 24), is convinced that everything in life can be explained rationally. Even his strange interest in sociobiology derives from the hope that "there might be algorithms to be found there which would crack the code of human behavior" (*FS* 25), and which would allow him to bridge the extreme divide between rationality and emotions that he is feeling. However, Frank's interest in sociobiology is weakened over the course of the sequels of the trilogy. This is mainly due to the epiphanic moment Frank experiences during a lunch-time lecture by the Khembalung delegation. While initially very skeptical and irritated upon listening to Drepung's and Rudra's elaborations on Buddhism and science, Frank begins to feel uncomfortably reminded of some of the thoughts he himself is having. He feels especially uncomfortable at the close of their talk, when the Khembalis critically question the value of scientific knowledge if one does not also think about to what end this knowledge can be used. Drepung and Rudra conclude their lecture by explaining that altruism—and thus practicing compassion—should lie at the core to scientific research. The Buddhists have dared to introduce morality and ethics into science. When Drepung adds, "An excess of reason is itself a form of madness" (*FS* 268), Frank feels as though he is being addressed personally, and starts to wonder about his own overreliance on scientific data. As he further contemplates this phrase, he comes to realize that "reason had never explained the existence of life in this universe" (*FS* 272) but what is done with knowledge, how one lived with it and made sense out of it, was also important. Furthermore, according to Buddhist practice, compassion is essential for humanity's survival. Frank realizes that at a time when the climate crisis had become so severe that it was threatening various peoples' livelihoods, science can no longer be solely occupied with empiricism, but needs to become part of the political decision-making process. This involves a radical rethinking, and Frank understands that emotion was needed—some kind of "*passionate reason*" (*FS* 275)—in really trying to answer what mattered in life. This moment proves to be a turning point in the series because it, ultimately, provokes a paradigm shift.

Instead of adhering to old assumptions and disciplinary structures, Frank realizes that a radical transformation is needed in the face of climate change. He withdraws his resignation in *Fifty Degrees Below*, and with the help of Diana Chang and Anna Quibler, begins to redirect the NSF to engage actively in mitigation efforts "for the sake of all the future generations of humanity" (*FS* 321). While presenting his new plan of action to the NSF board, Frank even draws on the Kuhnian idea of the paradigm shift and explains, "Kuhn postulated that in the usual state of affairs there is a general agreement to a group of core beliefs that structure people's theories, that's a paradigm, ... but anomalies crop up. Undeniable events occur that we can't cope with inside the old paradigm" (*FS* 322). Suggesting that the current climate crisis is such a moment when old paradigms no longer hold, Frank proposes that the NSF should act as the main driver of a paradigm shift, at the core of which lies the interaction of science with society and politics. The agency thus begins to engage overtly in politics, and even introduces a "scientific virtual candidate" in the upcoming election. Robinson hereby explicitly counters the division between science, politics, and society and illustrates that in times of climate change, a radical change in thinking about and communicating environmental problems is needed.

This paradigm shift is of importance for the plot, while also prompting Robinson to come up with new narrative techniques. In order to communicate climate change beyond disciplinary divides, Robinson breaks with genre conventions by not relying solely on science fiction and experimenting with a variety of narrative forms. Several times the author tellingly draws an intertextual reference to current discourses in climate science. Consistent with Roger Luckhurst's argument that the trilogy aims at confirming that climate change is real, Robinson closely adheres to the science of climate change while nevertheless teasing out its proleptic character by placing the story in the near future and leaving the reader in the dark about the origin of the scientific descriptions. The trilogy thereby epitomizes the "curious transitional space between fact and fiction" when it comes to climate change communication, in which "accumulated evidence and future modeling … research proposals, empirical finding[s], market report[s], science popularization[s], political strategy documents and science fiction [be]come very hard to distinguish from each other" (Luckhurst 172). While, interestingly, readers tend to believe that the scientific elaborations used in "realistic" fiction are accurate—sometimes mistakenly, as the Crichton example demonstrates—the science in science fiction is usually read as speculative (Csicsery-Ronay 114). Robinson's readers, however, knowing and valuing his works for their scientific accuracy, cannot comfortably rely on the belief that the science presented is fictive. Instead, by providing a seemingly accurate account and elaborate analysis of scientists' work, the novels constantly cast doubt on the distinction between fact and fiction, on what is certain or uncertain. Moreover, Robinson not only draws on climate science but also on Buddhist teachings as well as Emersonian nature writing. These intertextual references provide and emphasize a multi-perspectival approach to human-nature interactions in times of climate change.

Robinson deliberatively challenges readers' expectations in the climate series. Reading what is marketed as science fiction, the reader assumes to be immersed into a world of radical estrangement, but, nevertheless, expects plausible stories unfolding around a *novum*, "a rationally explicable material phenomenon, the result of an invention or discovery, whose unexpected appearance elicits a wholesale change in the perception of reality" (Csicsery-Ronay 5–6). Yet, unlike Robinson's *Mars* novels, in which an extraterrestrial planet introduces the estranging conditions, readers of the *Science in the Capital* trilogy find themselves in an all too-familiar, near-future setting. While Johns-Putra points out that this near-future setting allows for an "imaginative construction of climate change in a way that is psychologically and—one is compelled to add—politically and ideologically relevant" ("Ecocriticism" 754), she, similarly to Luckhurst, wonders if this makes Robinson's texts a *novum* or a work of science fiction at all. By drawing on science fiction critic Tom Moylan, who sees the most powerful potential of science fiction in its world-building impulse, Johns-Putra convincingly concludes that Robinson undermines the reader–science fiction text contract.[3] Rather than being drawn in by the narratological mega trope (Suvin's words) of the estranging *novum*, the reader recognizes the world that is built in Robinson's trilogy as her or his present one.[4] Several incidents in the novels resonate actual events in the history of Washington D.C., such as the floodings of Rock Creek Park, and leave no doubt about the real-world relationship.

Further blurring genre distinctions, Luckhurst argues, "Robinson brings to ecology the engineer paradigm that traditionally suffuse[s] American science fiction, but abandons the will to dominate or subdue nature often rehearsed there. Instead, the books seek compromise formations, mitigations that will restore an equilibrium between culture and nature" (179). Rather than restoring an equilibrium, Robinson, I would argue, maps out a whole new understanding of nature and culture, an understanding that resonates with Hulme's idea of "cosmopolitan climate," and that acknowledges that nature and culture are inseparably intertwined. In shifting toward this new paradigm, a network approach that goes beyond disciplinary boundaries is crucial. Thus Frank's first proposal at the NSF board meeting is to "to knit it all together" (FS 320) and to prioritize synergies. This requires the NSF to step out of its institutional business as usual and to not only engage with a different and more action-driven research approach, but also to secure cross-sectoral cooperation with politicians and military officials. Thinking about large-scale mitigation measures, Frank wonders, "What could be done if humanity were not trapped in its own institutions" (*SD* 245). Over the course of his trilogy,

Robinson explores this question and finally provides insight on the possibilities opened up by interdisciplinary work: successful carbon reduction policies, changes toward a more sustainable lifestyle, and international and cross-sectoral climate change mitigation strategies. Even though the narrative focuses on scientists most of the time, Robinson stresses that science alone cannot save the world. On the contrary, only by working together with politics and society can science and adaption to climate change be successful. As Luckhurst affirms, "To trace Frank is thus to trace out a limited but highly meaningful social totality, a matrix of power that fully intertwines science, society and politics" (175).

Thus, as the scientists in the series become increasingly involved in climate change mitigation efforts, the interaction among the different characters also becomes more complex and interwoven, thereby suggesting that a solution to the climate crisis can only derive from a network structure. I concur with Luckhurst, who likens Robinson's project to Latour's discussion of the workings of science, which is based on the idea that science only works through chains of connections and linkages. Drawing on Latour, Luckhurst further explains, "Science is therefore successful not to the degree that it isolates itself from society, but to the degree that it can be assessed by the 'number of points linked, the strength and length of the linkage, the nature of the obstacles'" (176). The importance of linkages is epitomized in the marriage between President Phil Chase and Frank's former boss, Diane Chang, who is now a science advisor to the president. This union of science and politics, as well as the already established marital bond between scientist Anna and political advisor Charlie, symbolically emphasize that climate change is a phenomenon of such complexity that it requires thinking beyond disciplinary boundaries and making a communal effort. Moreover, Robinson highlights the linkages and intricate connections between characters and the disciplinary approaches they stand for as well as between the characters and the spaces in which they move. Taking a closer look at his urban setting further reveals how human-nature boundaries are dissolved in the trilogy.

Urban Natures

Early environmental discourse with its dominant focus on nature writing based on ideas of wilderness or the pastoral rather fed into a nature-culture dichotomy. Constructing wilderness as antithetical to "unnatural civilization" (Cronon 80), as Cronon has convincingly argued, produced an image of nature that is only authentic when humans are absent from it.[5] With the rise of industrialization, the urban-nature divide was further deepened, "reinforced by the acculturated desire to keep nature both at a safe distance and within reach" (Buell, "Nature and City" 6). As Buell further elaborates, in the nineteenth century this dynamic was reflected in the designing of parks in American cities as much as in conservationists' efforts to create national parks. The tradition of propagating an urban-nature binary finally engendered the assumption in environmental discourse that "contact with the green world is more likely to develop eco-consciousness than life in the city" (Bennett 40).[6] As Andrew Ross concludes in an interview, "The literature of conservation—almost wholly devoted to nature worship in the 'cathedral of pines'—is persistent in its demonization of the city" ("The Social Claim" 16).

Further complicating the relationship between city and nature, environmental justice scholars have pointed out that environmental protectionism mainly emerged as part of the leisure activities of the white bourgeoisie. These scholars thereby criticize environmentalists' oblivion to environmental racism and seek to expose the correlation between toxic waste sites and marginalized communities.[7] By taking urban landscapes into consideration as nature, and by mapping out the political and socioeconomic structures of these landscapes, Michael Bennett finally introduced a new branch of ecocriticism, which he coined "social ecocriticism." According to Bennett, social ecocriticism is needed in order to "successfully incorporate urban environments within the domain of ecological criticism" (32). While Bennett still urged ecocritics to go beyond deep ecology and nature writing, this transformation has increasingly taken place in the past decade. As more than two-thirds of the US population lives in cities, ecocritics no longer neglect the built environments that people inhabit, and have engaged in a more diverse, fluid understanding of nature and culture. Environmental historian and writer Jenny Price most prominently called for a rethinking in writing about urban natures:

> In the past twenty-five years, the venerable American literature of nature writing has become distressingly marginal. Even my nature-loving and environmentalist friends tell me they never read it. Earnest, pious, and quite allergic to irony: none of these trademark qualities plays well in 2006. But to me, the core trouble is that nature writers have given us endless paeans to the wonders of wildness since Thoreau fled to Walden Pond, but need to tell us far more about our everyday lives in the places we actually live. ("Thirteen Ways of Seeing Nature in L.A.")

In her article, Price claims that in the city of Los Angeles, one can find as much nature as on any hiking trail as in national parks and points to questions that she believes are important to ask: how are these urban natures imagined, used, and transformed, by whom and to what end? Environmental humanities scholars have thus started to re-evaluate their research subject and to think beyond nature as something "out there." The connection between urban life and nature is also the focus of Brandt, Fluck, and Mehring's anthology on *Transcultural Spaces*. Significantly, in their introduction, these authors advocate that the contemporary environmental crisis can only be met by addressing "the modes of amalgamation and hybridization by which city and nature are constituted as complementary figures" (3).

Climate change fiction also prompts ecocritics to turn their attention to urban spaces and texts negotiating the urban-nature relationship. Here, cities are not only the backdrop against which climate change is played out—as dramatically depicted in the film *The Day After Tomorrow*, for example—but are also revealed to function as major contributors to CO_2 emissions.[8] British author Ian McEwan brings this notably to the fore when the protagonist of his novel *Solar*, a physicist and Nobel Prize winner now trying to reinvent solar energy, explains on one of his business flights:

> These days, whenever he came in over a big city he felt the same unease and fascination. The giant concrete wounds dressed with steel, these catheters of ceaseless traffic filing to and from the horizon—the remains of the natural world could only shrink before them. The pressure of numbers, the abundance of inventions, the blind forces of desires and needs looked unstoppable and were generating a heat, a modern kind of heat that had become, by clever shifts, his subject, his profession. The hot breath of civilization. (109)

Still, both of these works of fiction heavily rely on the urban-nature binary: the city as something defeated by nature and the city as the source of nature's destruction. They are nevertheless insightful because in mapping the changed nature of urban spaces, they bring to the fore the tropes that predominate our understanding of nature and the city.[9] Pointing to the importance of cultural texts for our understanding of climate change, geographer Harriet Bulkeley explains, "Understanding the urban nature of climate change requires that we examine not only the ways in which the issue is being produced, reconfigured and contested within cities but also how climate change has come to feature in the imagination and creation of urban futures" (4). While the trope of the city as an antagonist to nature continues to play an important role in the cultural imagination fed by a powerful tradition of anti-urban thought, as also illustrated in the examples above, Kim Stanley Robinson's *Science in the Capital* trilogy serves as a counterexample and helps to shed light on different conceptualizations of nature and urban existence. In fact, while recognizing the importance of taking cities into consideration in the discussion of the challenges that climate change poses, the trilogy, most importantly, exemplifies how the city is renegotiated as a hybrid space that is co-shaped by the natural and the cultural. In this space of climatecultures, humankind and nature are always co-constitutive.

Published only one year before Hurricane Katrina destroyed big parts of New Orleans, Robinson's *Forty Signs of Rain* deals with a major flood threatening to inundate another American city: Washington D.C. This first novel in the series, as well as the two sequels, powerfully argue against the idea that nature is something happening "out there" that can be watched from a safe distance. Instead, Robinson explores in detail the dramatic effects of climate change on the city and its inhabitants. While *The Day After Tomorrow*, which was coincidentally released the same year as the first part of Robinson's trilogy, depicts Washington's "defeat" by abrupt climate change, Robinson stages the urban-nature relationship in a more nuanced and complex manner. Instead of telling a story of urban decline or collapse, the trilogy explores how the city transformed by climate change becomes a space of possibility and innovation. Furthermore, Robinson illustrates that a response to climate change implies global systemic changes, and equally important, it requires a rethinking of smaller urban nature systems. Robinson hereby distances himself from a tradition of dystopias, which portray the city "in ruins," and employs a more utopian approach to catastrophe.[10]

Even though Robinson employs the idea of abrupt climate change, he refrains from using apocalyptic rhetoric and only very slowly builds up to the actual occurrence of extreme weather events. While the characters are engaged in issues related to climate change very early on in *Forty Signs of Rain*, it is only in the last 60 pages of the 400-page book that they are physically affected by it. Here, the reader finds out that storms have increased in frequency and a particularly ferocious front is now having detrimental effects on Southern California's coastline. Meanwhile on the East Coast, Washington D.C. is also facing two storms and a high tide.

Yet, even in describing how Charlie, Anna Quibler's husband, is forced to stay at his office because the streets of Washington are flooded, these passages do not read like part of an eco-thriller or a disaster novel. There is no huge tidal wave sweeping through the streets of the city killing everybody and destroying everything in its path. On the contrary, Robinson describes in detail and in a rather neutral and almost report-like manner the transformation of urban landscapes. When Charlie looks outside his office window and sees a "City floored by water" (*FS* 361), he is shocked but at the same time fascinated. Observing how the urban space surrounding him becomes inundated, he still remains calm because he feels safe in his high-rise office building. He joins his colleagues who are watching the news in order to know more about the flooding situation. Charlie himself points to the paradoxical nature of this event: "It was strange to see how they were directly involved in an obviously historical moment, right in the middle of it in fact, and yet they too were watching it on TV" (*FS* 368).

While the reference to news images reinforces Robinson's "objective" and report-like narrative tone, it also critically points to the power of the media in shaping our experience of catastrophe. Yet, Charlie's observation also allows for a shift in narrative perspective. Zooming out of Charlie's spatially limited range, the camera images provide insights from all over the city and reveal the large-scale ramifications of the flooding. As Charlie and his colleagues skip from TV channel to TV channel, a comprehensive overview is gained, which shows how the city as a whole has been affected. At the same time, the detailed urban descriptions allow readers to reassemble the differently affected parts of the city and even readers unfamiliar with Washington D.C. to envision how the city has become submerged by water. Yet, leaving only short time for recognition, the author is quick to defamiliarize the city's landmarks, describing how the famous Mall has been turned into a lake with the water rising "just up to the steps of the White House and the Capitol," and even the Lincoln Memorial "despite its pedestal mound, appeared to be flooded up to about Lincoln's feet" (FS 365). As the water threatens to submerge key icons of American politics and power, the severity of the situation is emphasized. More importantly, the cityscape itself is reconfigured as nature "reassert[ing] itself even within the privileged and symbolic landscape of the nation's capital" (Kilgore 5), and highlighting that culture and nature are not separate entities.

By illustrating how climate change affects urban spaces, Robinson simultaneously reveals how the power structures of the city are inscribed in its built form. Urban sociologist, Saskia Sassen, has most notably argued, "how power and the lack of power inscribe themselves in the urban landscape" (2), drawing on the corporate complex and the immigrant community as examples. Taking into consideration how these places are created in concurrence with topographical features of the city seems to be a valuable addition to her argument. In *Forty Signs of Rain*, Robinson acknowledges in the passages about the flooding that not all of the city's population experiences disaster equally, that some may be more vulnerable to "nature." While the Capitol and the White House are conveniently located on small elevations, poorer housing areas in the flat areas of the Anacostia River valley are heavily affected. Charlie critically reflects on the "usual correlation of money and elevation" (*FS* 367) when he realizes that "snobbish little Georgetown" is located on a rise. He is, nevertheless, also relieved that his wife and children are safe in their house in Bethesda.

The sequel *Fifty Degrees Below*, in depicting how Washington's inhabitants have to endure a particularly harsh winter, also reveals how social inequalities are aggravated by climate change. The author shows that while Charlie's family and other characters like Frank have the financial means to "comfortably" endure these weather extremes, there are other people, such as the group of homeless guys that Frank regularly visits at their campsite in Rock Creek Park, who are badly suffering from the changed climatic conditions. Several passages illustrate these social inequalities: the Quibler family, for example, reacts to a power outage when it is fifty degrees below zero outside by coming together inside their house and enjoying the fireplace and moment of family time, while the group of homeless people is forced to huddle together at the bonfire in order not to freeze to death. Furthermore, Frank, who is well-equipped with his polar gear and enjoys going hiking in the park even in the cold, arrogantly mocks the people who are still running around wearing cotton. While he tries to help his homeless friends by giving them advice on clothing and on restructuring their campsite in order to stay warm, he cannot understand why they are so unwilling to move into one of the city's shelters. After the mass floods in the novel, however, Frank himself begins living in a tree house in Rock Creek Park and in his car for several months. When asked why he hasn't moved indoors yet, he replies "I don't need to" (*FD* 424), making clear that he is a homeless person by choice.

Even though the trilogy portrays how climate change leads to deteriorating living conditions and exacerbates social inequalities, it simultaneously postulates that the moment of climatic crisis also holds the possibility of overcoming differences. "Nature" is thus also portrayed as having the power to change socio-cultural patterns built in concrete. In fact, the flood and later the cold spell transform Washington to such an extent that they destroy various urban structures and allow for the emergence of new communal spaces. This is foreshadowed in the description of Frank's experience of the flood. Since Frank is at the NSF office when the flooding occurs, which is located in the not-yet-flooded areas on the other side of the river, he is still able leave the building and joins a group of volunteers helping to build a sandbag wall to protect the Arlington National Cemetery from inundation. Similar to his epiphanic moment at the Buddhist lecture, this moment proves to be another turning point for Frank, who is no longer a misanthrope and now enjoys the moment of communal effort in trying to protect the city. In working toward a common goal, "He felt deeply happy, and looking around he could see that everyone else was happy too. That's what happens, he thought, … It takes something like this to free people to be always generous" (*FS* 374). Robinson demonstrates in this passage that crisis does not necessarily lead to paralysis and may even encourage solidarity. Even though the city is inundated, there is a utopian moment within this catastrophe. This also explains why Robinson does not revert to apocalyptic rhetoric in his depiction of the flood. Instead, the trilogy is, as he himself asserts, "a domestic comedy about a global catastrophe—and how we might avoid it, or even counter it once it starts" ("Amazon Shorts" 17). He continues to explain that disaster, though destructive, also bears the opportunity for change and that "Depending how we react to it, the possibility of abrupt climate change could be a *good thing*" (ibid.). Refraining from staging the flood as a battle against nature, Robinson instead posits it as a positive renegotiation of urban space, which instead of anomie produces spaces of encounter and exchange.

In the novel, the possibilities of change that a positive and more cosmopolitan urban-nature framing provide are juxtaposed with the binary nature-culture rhetoric that dominates conservative right-wing American politics. In the beginning of the second book of the trilogy, *Fifty Degrees Below*, the author critically exposes this rhetoric, which fails to acknowledge the opportunities that result from a reconfiguration of cityscape. As a reaction to the flood or in his words to "this act of climactic terrorism," the president announces that "From now on … we are at a state of war with nature. We will work until we have made this city even more like it was than before" (*FD* 4, sic). Tellingly, while the president still hopes to return to some former pre–climate change state, Phil Chase, who at the end of *Fifty Degrees Below* is newly elected president, already cruises over the Potomac with the "happy look of someone who had already lit out for the territory" (*FS* 393). And after the city is also hit by the extremely severe cold, even the right-wing president has to realize that some changes, such as global warming, are not reversible. Yet, Robinson's depiction of the transformation of the city by the cold is also subtle. This description of the dangers and problems of such a dramatic drop in temperature lacks sensational brutality due to Robinson's slow narrative—as with the floods, the cold wave occurs only in the last third of the second book. More importantly, the disaster is again portrayed in a nuanced manner: the cold not only causes problems, but also joy, as the description of the characters' manifold outdoor activities in the snow demonstrates.

Furthermore, the trilogy illustrates here again how climate change, in *Fifty Degrees Below* exemplified by the extreme cold, challenges former city structures and thereby produces new contact zones.[11] This underscores Lefebvre's famous work on the "production of space," which acknowledges that space is a dynamic process rather than a static category and cannot be separated from social interactions. As Derek Gregory and John Urry elaborate in a publication that draws together human geography and sociology, space is "not merely an arena in which social life unfolds but a medium through which social life is produced and reproduced" (3). This interrelation between spatial structures and social life is most prominently brought to fore in the trilogy when Frank describes the scenery that unfolds in front of his eyes on the Potomac River:

> Out here on the river you could finally see that [most of the Washington D.C. population was black] in a way that Northwest and Arlington never revealed. It was like Carnavale on ice, the celebrants improvising clothing that was warm enough to keep them out there, which then became costumes too. A giant steel drum band added to the Caribbean flavor. Snowfights and slip-and-slides, break dancing and curling that was more like bowling, touch football, tackle football, it was all happening out there between Virginia and the District, on this sudden new terrain ... It seemed to him to be an extraordinarily beautiful populace, every race and ethnicity on Earth represented ... all partying together on the frozen Potomac, until with the dark it got too cold to stay out any longer. (*FD* 421–2, sic)

The trilogy adds an important aspect to the discussion of the production of space by pointing to the influence of environmental processes on built spaces. In exploring how climate change is shaping urban spaces anew, Robinson depicts how contact zones emerge, which bring together different social and ethnic groups. He thus goes beyond declensionist environmentalist rhetoric and rather than lamenting what has been lost, he highlights what is gained.

While Robinson has been criticized for his employment of a predominantly white character ensemble, this passage acknowledges Washington's diverse ethnic makeup. Passages like the one above also illustrate the author's attempts to envision how abrupt disaster may eliminate previous spatial segregation. The narrator here presents a vision that counters the Republican president's nature rhetoric based on "patriotism as xenophobia" (*FD* 546). "Nature's nation" is no longer white and Anglo-American, but cosmopolitan. As Frank welcomes the city's diversity, this microcosmic vision is also projected onto the global scale since Phil Chase's presidential campaign for change builds on a vision in which "everyone was part of it" (ibid.). In one of his earlier speeches, the soon-to-be president Chase even overtly calls on his fellow citizens to "join the effort to invent a global civilization" (*FD* 261).[12] The importance of thinking beyond national borders in the face of disaster is further emphasized when the flood threatens to inundate Washington's zoo. The Khembalis, "experts in flood response" (*FS* 380), urge staff members to evacuate the zoo and to release the animals according to plans by the Global Disaster Information Network, "which had a complete protocol for this very scenario" (ibid.). The global here serves as a source of information to deal with the local. While the narrative often emphasizes the importance of thinking globally and beyond ethnic or national boundaries in several instances, the city remains at the center of the narration. Yet, the manifold references to the global emphasize that Washington D.C. only serves to illustrate on a micro level what Robinson postulates as a planetary response to climate change.

Returning to the analysis of the transformation of Washington's urban spaces by abrupt climate change, parks—especially Rock Creek Park—take on a central position, because they epitomize the nature-culture amalgamation in the city. Most of the parks established in major American cities in the nineteenth century were part of a moralizing effort "intended to take the rough edge off the immigrant soul" (Ross, "The Social Claim" 22). The park was thus a clearly demarcated space in the city, and contact with its "domesticated" nature was considered to have "civilizing" effects. Furthermore, following European examples, the park served as a pastoral sanctuary inside the urban jungle. Today, parks are often some of the few remaining public spaces in the city and, as sites of "intense socialization" (ibid.), are highly contested terrain. Disputes thus often revolve around access to parks rather than about the interactions between humans and nature inside the park.

Robinson's trilogy brings attention to this kind of interaction between humans and nature. His novels explore how climate change complicates ideas of the park as a space set apart from the city and instead presents it as space of close entanglement. It is furthermore the prime space of the narrative, which allows teasing out the multifaceted human-nature relationship. The flood that hits Washington alters the landscape of Rock Creek Park, and also changes its entire ecosystem, which now has been turned into a new habitat for several zoo animals. At first, it seems that the park is no longer a domesticated sphere, but is claimed by wilderness. Yet, while there are feral animals roaming the park, and harsh weather conditions have come to be the norm, Robinson provides an image of wilderness that does not rely on the absence of the human. Instead of uninhabited wilderness, the park figures as a "middle landscape between the urban and the wild" (Kilgore 5). And even more than a middle landscape, the park functions as a hybrid space—a space that is neither only natural nor only cultural/human/urban. Frank's life in the tree house further underscores this reading. Since Frank has problems finding a new apartment after the flood, he ultimately decides to live in the park and to build a tree house there. However, as Cho has explained, this does not figure as the return of man to untouched nature because, "[Frank's] wilderness is a product of global warming and scientific experimentation [Frank's attempt to adapt to a new lifestyle], both of which stem from human activity, and in that sense, it is postnatural" (Cho 43). The ambivalent "nature" of climate change is thus epitomized, as Frank enters a park that was turned into wilderness by climate change, but since climate change is anthropogenic, it poses the question if the park can really be wilderness as commonly understood. Finally, the portrayal of the dynamic nature of the park serves to further illustrate how climatecultures change human-nature relationships.

Shortly after entering the forest, Frank becomes acquainted with three different groups of people also present in the park. First, Frank meets a group of homeless people, the "bros," who have reclaimed their home by setting up a campsite in the park. In fact, this campsite, at once refuge and place of danger, figures as an important contact zone throughout the trilogy, in which urban nature is constantly renegotiated. For instance, while Frank first comes to the park in the hope of finding solitude and spiritual renewal, he instead stumbles across the campsite and the bros and is quickly reminded that the park, albeit now officially off-limits to people, is not an uninhabited space. Instead, it is a space where the problems of the city—unaffordable housing, a dysfunctional social system, and exclusion—are most noticeably brought to the fore. Yet it is also the space where societal rules do not apply and, as such, it provides a certain freedom.

During one of his visits of the bros, Frank meets the second community roaming the park: a group of Ultimate Frisbee players. As Frank begins to play with them, he finds out that they are fregans. Exposing the paradoxes of contemporary "throwaway society," fregans live on food from garbage containers and sleep in abandoned houses. In their life off the grid, they constantly move between park and urban space, but the park mainly figures as a place for amusement outside of societal norms. Finally, Frank explores Rock Creek Park with a third group, the Feral Observation Group (FOG), which forms part of the Friends of the National Zoo. While the majority of the released zoo animals were recaptured again after the flood, many still remain in the park and are—though closely monitored by FOG—supposed to remain free. For this group of people, who remain observers, the park does indeed figure as wilderness. Yet as the narrative traces Frank, who moves between the different communities, the various meanings of the park are also renegotiated, stressing that it is neither wild nor domesticated. Robinson thereby refuses to give one dominant interpretation, and instead depicts the park as a site of refuge for humans and animals alike, as a place of encounter and interaction, a place of danger (human and non-human), but also of quiet and solitude. It is ultimately a space that forecloses being categorized as just natural or just cultural/urban space.

The depiction of Frank's transformation into "Alpine man" serves to further emphasize humankind's embeddedness in nature in a climatically changed future. Frank functions as the key character for the observation of how the transformed city speaks and interacts with its inhabitants. Similar to the *flâneur*, a central figure in studies about modern urban life, Frank walks through the city and acts as a social commentator in observing the city life around him. As most notably described by Charles Baudelaire and Walter Benjamin, the *flâneur* views the streets as the most vital field of observation, yet he rarely interacts with others and often remains invisible to the crowd. This also holds true for Frank, who wanders the streets of Washington D.C. and provides the reader with commentaries about contemporary American society in general as well as observations of city life in particular, fused with his strange interest in sociobiology. Furthermore, as Adeline Johns-Putra argues, the detail in geographical reconstruction provided in Frank's accounts invites readers to immerse in the cityscapes described and therefore to participate in Robinson's terraforming project (756). In his article "Imagining Abrupt Climate Change," Robinson explains that the term terraforming was introduced in the science fiction of the 1930s by Jack Williamson and used to describe the engineering effort to make other planets in space similarly inhabitable like the Earth.[13] Yet, while working on his most well-known *Mars* trilogy, Robinson realized that instead of other planets "we were already terraforming Earth, in the here and now, but by accident, and in ignorance of how it worked or what might happen" (1). The *Science in the Capital* trilogy is then the result of his closer engagement with this kind of terraforming. Patrick Murphy has added that in depicting how humankind is already transforming the planet, Robinson also highlights "ethical questions of engineering the biosphere and engineering the species in the context of human and environmental sustainability" ("Engineering Planets" 54). More importantly, Robinson shows the deep interwovenness of human and environment in times of climate change.

So, when the city is affected by climate change, walking the streets of Washington as a detached observer is no longer possible. Instead, Frank turns from *flâneur* into "Alpine man" (*FD* 429). When temperatures begin to fall, Frank buys special polar gear in order to continue wandering outdoors and describes how he "could not be more set; he was probably the best-dressed man in the city" (ibid.). Yet, "Alpine man" does not solely designate a change in style or outer appearance. Instead, it constitutes one part of Frank's "parceled" life—as he explains his attempts to constantly renegotiate his identity anew. In preparing for extreme weather conditions, Frank seeks out his Paleolithic self while being reminded of Ötzi, "who had died crossing a Tyrolean pass some five thousand years ago" (*FD* 347) and whose equipment, which was preserved in ice over the years, bears a lot of resemblance to his own equipment. Kilgore argues that the parallels between Frank and Ötzi, the "original" Alpine man, not only serve to "imagine a productive escape from modern limitations and restrictions" but that "the gesture also reveals a continuity linking the present to several millennia of human endeavor" (6). So Frank's transformation into Alpine man epitomizes humanity's intricate and long-term involvement in nature. However, rather than escaping from "modernity" to "nature," as Kilgore argues, Frank as Alpine man embodies the connection of the two.

Furthermore, as Alpine man, Frank is able to transcend boundaries by moving back and forth between the city and the park and thereby reveals not only the people's interaction but also climate change's interaction with the city. Notably, Frank is no longer the detached, *flâneur*-like observer, but engages with city inhabitants and nature alike. As the reader follows the routes that Alpine man takes, it becomes clear that human and non-human, culture and nature communally play a role in shaping and renegotiating urban spaces. Yet, by closely mapping the transformation of urban space, Robinson also invites readers to critically examine their role in terraforming. His trilogy in a way exemplifies Jill Bennett's description of the Anthropocene: "The entire world population suddenly finds itself part of this "cene," [and this constitutes] a critical factor in determining how it will play out." Providing a narrative about how humans may shape these "-cenes" to a positive outcome ultimately shapes Robinson's utopian writing project.

Permaculture in the Anthropocene

Even though the Anthropocene, this new geological era proclaimed by chemist Paul Crutzen, remains contested and geologists are still investigating whether human action has impacted the planet to such an extent that we have indeed substituted the Holocene with a new man-made geological era, the introduction of this term for the analysis of Robinson's work is useful. The employment of "Anthropocene" most notably acknowledges that the human and the non-human can no longer be thought of independently from one another in planetary history. Considering the wide range of academic disciplines that engage with the term further illustrates that it is as much a geophysical as well as a social and discursive phenomenon. The discussions about this radical transformation of the human-nature relationship are informed by two different assumptions about human agency (Heise, "Comparative Ecocriticism" 17). On the one hand, Crutzen and environmentalists such as Bill McKibben, for example, engage in declensionist rhetoric and warn that humankind's destructive force has led to irrevocable changes in the natural environment. Humankind's interaction with nature, according to this line of argument, thus needs to be reduced to a minimum. On the other hand, geoengineers and others have drawn on the notion of the Anthropocene to promote an interventionist and rather optimistic approach based on the belief that humanity's ingenuity will serve as the basis upon which to reshape the nature of the future planet.[14] However, climate change as the materialization of the unintended consequences of our attempt to dominate nature highlights human's incapability of controlling any such intervention.

Common to both of these arguments, and inherent to the term Anthropocene itself, is to put the human at the center of the narration. However, Chakrabarty tellingly wonders: who are these humans that now communally as a species make up a geological force?[15] And even if a transformation of such planetary magnitude requires a re-conceptualization of human collective agency in terms of species thinking, how can one experience oneself as a species? Chakrabarty concludes that

> We humans never experience ourselves as a species. We can only intellectually comprehend or infer the existence of the human species but never experience it as such. There could be no phenomenology of us as a species. Even if we were to emotionally identify with a word like *mankind*, we would not know what being a species is, for, in species history, humans are only an instance of the concept species as indeed would be any other life form. But one never experiences being a concept. ("The Climate of History" 220)

While Chakrabarty claims the impossibility of grasping one's own agency as a species in an unmediated manner, Heise convincingly points out in response to his argument, that abstract categories can nevertheless be experienced. She continues explaining that while conceiving of species as a social category is difficult because it requires an awareness of human's ecological embeddedness, it is, nevertheless, possible "by means of its own set of rhetorical, symbolic, legal, and institutional structures" (Heise, "Comparative Ecocriticism" 20). Teasing out those structures may be a valuable contribution from the field of ecocriticism to ongoing discussions about the Anthropocene (ibid.). Turning to Robinson's trilogy proves to be valuable in this context because it "fosters new, expansive visions of humankind's co-implication in the natural world" (Markley 2). Markley concludes that Robinson's new vision unsettles ideas of nature based on harmony and balance. In addition to promoting a dynamic understanding of nature, the trilogy also reflects on what it means to be human in the age of the Anthropocene. It thereby explores how humankind is terraforming the planet, and, at the same time shows how this is changing us as a species.[16] Yet, the human constitutes here neither the center of attention nor the dominant force of narration. Instead, Robinson offers a climateculture perspective that acknowledges human existence as part of multiple natureculture assemblages. The trilogy then gives insight on the idea of the human in an age of the Anthropocene that also Chakrabarty has called for: it is a view of the human "as a geophysical force and as a political agent, as a bearer of rights and as author of actions; subject to both the stochastic forces of nature (being itself one such force collectively) and open to the contingency of individual human experience" (Chakrabarty, "Postcolonial Studies" 14).

In reimagining this changing human-nature relationship, Robinson navigates between the two underlying assumptions of human agency discussed above. On the one hand, he draws on the interventionist approach: by introducing large-scale geoengineering as a way to mitigate climate change, he explores ways to responsibly terraform the planet.[17] Certainly, Robinson's engagement in his writing with geoengineering projects complies with the reading expectations of his science fiction audience and could be interpreted as human hubris in trying to dominate nature. Yet, his scientist characters caution that the long-term consequences of interventions such as carbon sequestration with biogenetically modified lichen or enhancing the salt level to restart the North Atlantic current are unknown. Moreover, the story reveals that exploitative capitalist practices and past interventions into the Earth's system have been the cause of the current climate crisis. However, as a writer of utopian science fiction, Robinson is not so much interested in condemning actions of the past or lamenting what has been lost.[18] In *Sixty Days and Counting* President Chase tellingly proclaims on his blog: "No more blaming the past" (*SD* 357). Instead, Robinson explores in detail how society takes the moment of climate crisis as a chance to build something new and to move forward in spite of catastrophe (Cho 33), even if this requires embracing geoengineering as a way to mitigate climate change.

Still, the author highlights that sustainably terraforming the planet not only involves a change in the geophysical make up of the planet, but also that it encompasses a reflection of the underlying value and belief system on which our human development depends. The trilogy thereby explores what we as humans can technologically do to "solve" the climate change problem, while also depicting how the very idea of climate change reshapes the ways in which we think about our existence and our collective interaction on Earth. Again, turning to Phil Chase's presidential blog entries helps to illustrate this. Chase here explains, "We'll share the world with all the other creatures. It will be an ongoing project that will never end" (*SD* 516). The capitalist system based on the maximization of one's own affluence, on accumulation and growth, according to the President in the trilogy, fails to acknowledge that humans are only one species coexisting among many others. Instead of selfishly exploiting Earth, he calls for respect toward the non-human. Moreover, Chase proposes an understanding of life on Earth that acknowledges that nature is always in flux and thus proposes to embrace change and constant adaptation. Yet, this approach does not free humankind from an active responsibility toward life on Earth. On the contrary, Chase emphasizes that since it will never be possible to dominate nature, humans need to acknowledge nature's agency and begin to see themselves as responsible actors within a planetary network system.

Instead of dwelling on insuperable differences between nature and culture, the trilogy provides a means of seeing ourselves as part of climatecultures. This is further brought to the fore in the description of Frank's attempt to live according to optimodality. In living optimodally, Frank transcends urban-nature boundaries, as already indicated above, as part of his attempt to seek out a parceled life. This involves "moving from one thing to another, among differing populations; behaving as if a different person in each situation" (*FD* 88). Frank actively engages in his new life of optimodality and soon turns from a solitary scientist into a person who moves between different groups, activities, disciplines, and ideas. He participates in spiritual debates with Rudra, a Buddhist from the Khembali delegation, discusses the financial aspects of large scale geoengineering with US military officers, fights over the accuracy of scientific data with Anna Quibler, muses over quotes by Emerson and Thoreau with the fregans, and enjoys his frequent canoe trips on the Potomac River. In following the different activities that Frank pursues in the aftermath of abrupt climate change, the reader is soon made aware of the various layers of human-nature interaction. The diverse stories about Frank's encounters with nature while living optimodally then serve as a way for readers to imagine nature in its various shapes and to draw different associations of the human relationship to these changing natures. Not only does Frank enthusiastically read Emerson and Thoreau throughout the trilogy, but he also soon encourages people, including Diana Chang and even the president, to read the transcendentalists' texts as well. The epiphanic nature described by the Emerson and Thoreau is juxtaposed with the "catastrophic" nature that the characters encounter, with the biotechnical nature that the NSF members engage with, and again with the different sacred nature that the Khembali delegation describes. All these different natures come to influence the characters in the trilogy and it is only through a constant rethinking of their own subject position in relation to nature that change toward a more sustainable environment is made possible.

In depicting how humans and the environment mutually constitute one another, the *Science in the Capital* trilogy also explores humankind's complicity in events that many contemporary Western societies would not necessarily connect to themselves. The complete inundation of the island of Khembalung represents the threat of rising sea levels due to climate change that many island nations face. However, Robinson refrains from moralizing rhetoric. Even while portraying the tragedy of the Khembalis' loss, he does not leave much time in the narration for mourning. Instead, he foregrounds the Buddhist belief in rebirth and spiritual renewal and depicts how the Khembalis embrace change and build a new home in the United States. This forms part of the trilogy's utopian impulse to present alternative ways of living and interacting with nature after abrupt climate change disaster has occurred. Again, Robinson emphasizes that while humans have a wide-ranging impact on the transformation of the planet, this change is not unidirectional and may also prompt a change in our own species understanding. Impressed by the Khembalis' calm reaction to the loss of their home, and after engaging in various discussions with them, Frank and the Quibler family also find a way of dealing with crisis through an approach based on Buddhism, practicing compassion, and a more holistic understanding of nature.

This spiritual turn also requires the courage to accept the unknown that lies ahead and to step out of one's comfort zone in order to adapt to a new situation. In the beginning of the third book of the trilogy, Charlie describes his fear of change because it involves becoming: "And so: change. The inexorable emergence of difference in time. Becoming. One of the fundamental mysteries" (*SD* 48). However, remembering the Buddhists' basics, he finally admits that change "had to be lived, and then the next moment embraced as it arrived" (*SD* 49). This passage illustrates how Buddhist ideas about life help Charlie to engage in action instead of being paralyzed by crisis. Embracing new ideas about nature and acknowledging nature's constant transformation is a crucial part in Robinson's project to illuminate how to move forward in a world affected by climate change.

As the characters come to recognize the complexity of nature and the various layers of the human-nature interaction, they also begin to think of alternatives to their current lifestyles and try to work toward more sustainable practices. A great part of the trilogy focuses on small changes in the characters' daily routines and their individual but continuous efforts to adapt to constantly changing natures. Thus, the Quibler family, who lives in a typical American suburb, decides to redesign their backyard into garden beds and happily proclaims, "No more suburban lawn wasting their yard space!" (*SD* 249). In addition, they develop several routines for saving more energy and reducing their carbon footprint. Portraying in detail how the family juggles work, taking care of their two children, coping with the blackouts and food hoarding during the extreme weather events, and still manages to live more sustainably, Robinson illustrates that while change is difficult, it is possible. The author also explores a more radical change of lifestyle by tracing the group of fregans. Frank joins them on several occasions during one of their fregan potlucks. Every time he is astonished by the diversity of the group. Young and old, and from different ethnic groups, these fregans come together in abandoned houses and share their food. While this is the most extreme "set of habits Frank had lived so far" (*SD* 355), he wonders if this was the way people should live and googles "to see if any studies had been done to determine how quickly new habits were internalized as a norm" (ibid.). While ultimately Frank only joins the group temporarily, Robinson's depiction of the fregan way of a life serves as one, though the most radical, example in the trilogy of re-envisioning the nature-human interaction with a greater eye toward sustainable behavior.

While individual actions play an important role, Robinson also presents a vision of a wider political and societal change over the course of the trilogy. Most notably, Frank's left-oriented Argentinean colleague Edgardo, but also president Phil Chase, aim for what they call "permaculture." Their idea is similar to the concept of permaculture first introduced by ecologist Bill Mollison, who described it as "a philosophy of working with, rather than against nature; of protracted & thoughtful observation rather than protracted & thoughtless labour; & of looking at plants & animals in all their functions, rather than treating any area as a single-product system" (n.p., sic). Mollinson alludes in this definition to the importance of acknowledging the dynamics of nature and stresses the importance of taking into consideration the interaction of all the different elements in a particular ecosystem. In the trilogy, President Chase also clarifies that the prefix "perma" does not refer to a static state. On the contrary, he explains, "By permaculture I mean a culture that can be sustained permanently. Not unchanging, that's impossible, we have to stay dynamic, because conditions will change, and we will have to adapt to those new conditions, and continue to try to make things even better" (*SD* 517). In *Fifty Degrees Below*, Edgardo gives a more concrete account of what such a permaculture entails: From clean technologies and protection of human welfare to the acknowledgement that all components of the biosphere, which includes humans, are interdependent, to the more equal distribution of wealth. Robinson implicitly here provides his vision of a more just, equal, and sustainable future. While the extreme weather events that occur throughout the novels serve as a catalyst for this societal change, Robinson stresses that in their aftermath, persistent transformation is required. In addition, the sheer length of the trilogy seems to underscore the characters' continuous and long-lasting efforts in working toward a new way of life.

The trilogy's portrayal of the difficulties involved in this transformation shows that Robinson depicts a "utopia in the making." Contrary to the description of a perfect future society, Robinson's near-future setting and slow narrative mode stress the resemblance of the fictional world to real world of the reader. As Johns-Putra and Trexler have explained, "The reader does not look upon utopia as a setting, but experiences utopia in and as progress" (Trexler and Johns-Putra 191). The trilogy thus serves as an example of the author's attempt of "imagineering," which I have described elsewhere as creating a manual for critical intervention derived from creative ideas (Mehnert 31). Illustrating how his characters slowly begin to change and adapt to new ideas and lifestyles, Robinson encourages his readers to do the same. Gib Prettyman's analysis of Robinson's trilogy also argues for the close connection between Robinson's writing project and his engagement with "eco-activism." Prettyman draws an analogy with biotechnology to describe Robinson's usage of genres. By paralleling genes and genres, he points to the encodings that determine our behavior. He explains, "Through a series of interwoven narrative tropes … Robinson draws attention to the genetic function of genres and explores contemporary possibilities for modifying them" (182). According to Prettyman, elements of various genres—comedy, thriller, Robinsonade, science fiction, and utopia—function as disruptions to cultural norms and thus also to established patterns of behavior. As a consequence, the reader has to constantly modify his own subject position, and, similar to the characters in the trilogy, has to adapt to continuous reconfigurations—a necessary aspect of Robinson's larger and more complex utopian project of adapting to and mitigating climate change.

The socio-political intention of Robinson's writing is not to be neglected. Robinson has repeatedly pointed out that he considers literature to be part of a political commitment. To him "symbolic acts are also real acts" (Szeman and Whiteman 182). In *Forty Signs of Rain* he makes this very explicit, when Frank explains that language is like gene expression: "From an instruction sequence, like a gene, to what the instruction creates. Language to thought. Or to meaning, or comprehension. Whatever! To some kind of living thought" (*FS* 292). By telling the story of adaption to climate change as utopian science fiction, Robinson ultimately offers imaginary templates for "actively living in the Anthropocene" (Bennett). He thus explores alternatives to current lifestyles, proposes new coalitions and collaborations, and provides a new sense of the future, beyond the apocalyptic gloom-and-doom narrative.

Still, the utopian impulse in Robinson's trilogy does not free humankind from its responsibility for causing the climate crisis in the first place. On the contrary, the employment of abrupt climate change as a motif is crucial because it allows for the reflection of ourselves as a species that is having a major impact on the planet we inhabit. It points to our responsibility not by turning to the past and condemning polluting actions that caused climate change, but by stressing the current need to change. As daily routines to which humans have grown accustomed are interrupted, urban landscapes are remapped, disciplinary confines broken, and belief systems questioned, the moment allows for new ways of envisioning life on Earth. As Phil Chase summarizes in the closing sentences of *Sixty Days and Counting*: "Our time has to be understood as a narrow gate, a window of opportunity, a crux point in history. It's the moment when we took responsibility for life on Earth" (517).

Finally, instead of perpetuating an apocalyptic scenario where humankind is punished by nature—as happens in many literary portrayals—Robinson reminds us that we are deeply embedded in nature. The trilogy thus serves here as an example of the reconceptualization of the representation of nature. As Feldman and Hsu claim, taking as a starting point the "mutual imbrication of nature and society ... demands that we rethink the very concept of representation: nature cannot be represented as something separate and complete in itself, but rather must be seen in relation to socially differentiated bodies" (Feldman and Hsu 206). In their article, the critics ultimately emphasize the importance of considering the social dimensions of environmental problems. Robinson's trilogy goes even further. In addition to depicting the social dimension, he also explores nature in relation to politics, economy, science, religion, values, and beliefs. If we become aware of our multiple interactions with various natures, we might also think of ways to responsibly terraform the planet. Markley also affirms that the trilogy ultimately conveys that it is "only through a collective re-thinking of history, science and Nature that a new civilization can begin to emerge" (2, sic).

Robinson's trilogy exemplifies new ways of imagining and describing climate change that do not rely on negative environmentalist rhetoric. He does so by providing insights on the multiple facets of nature and humankind's co-implication in it. "Nature" is here neither solely catastrophic nor solely pastoral, but part of a complex and constantly changing matrix. It is the biological materiality "out there" as well as a discursive category. By depicting a climatically changed future in detail, Robinson complicates nature-culture dichotomies by foregrounding a hybrid and co-produced nature. The human is thereby always already a part of that nature. The importance of interdisciplinary network structures, the remapping of urban natures, and the introduction of "permaculture" prominently exemplify Robinson's emphasis on climatecultures in his trilogy. Envisioning climate change futures beyond the nature-culture divide, moreover, allows Robinson to elaborate on human agency in a way that does not rely on the oppositional rhetoric alluded to by Heise. He refrains from a declensionist narrative that focuses on teasing out humanity's guilt for ecological crisis, and also avoids establishing humankind's dominion over nature by endorsing a techno-managerial fix. Instead, Robinson postulates that there is no pristine nature to be preserved, nor a wild nature to be dominated, but only a multitude of climatecultures. A response to the climate crisis, accordingly, must then be based not only on a wider conception of what nature might mean and thus needs to engage with questions such as: what kinds of nature do we as a species want to create? What kind of natures ensure a just life on the planet? And what losses of natures do we possibly need to accept for a future life on the planet? Robinson's utopian science fiction trilogy provides one possible story of adaptation that engages these questions, and illustrates how responsible terraforming and a broad scientific-political coalition, as much as individual sustainable action, function as ways to deal with the climate crisis.

Notes

1. While Hulme's employment of the term "cosmopolitanism" is plausible in the context of his argument and within his disciplinary, and thus geographical, argumentative framework, literary and cultural

studies scholars may criticize its usage here as overly simplistic. As the discussion in Chap. "Scaling Climate Change: The Transformation of Place in Climate Change Fiction" has shown, the term emerged in the context of globalization studies and has been discussed by a variety of scholars. Hulme, following Beck's ideas of cosmopolitanism, points to the positive connotations adhering to cosmopolitanism. Critics, however, have argued that cosmopolitanism is an elitist term that celebrates the dissolution of boundaries while neglecting that processes of detachment from place often happen involuntarily and are thus very painful experiences. As Heise argues, Beck's usage of the term "cosmopolitanism" falls short of including the multilayered discussions in anthropology, philosophy, and cultural studies ("Cultures of Risk" 21), and its universalist employment seems oblivious to differences of class and gender. While being aware of the problematic terminology, I will use Hulme's term "cosmopolitan" here in order to refer to the hybrid nature of climate change.

2. There is an elaborate explanation of the term "terraforming" in my discussion of urban natures.
3. Moylan explains, that readers engaging in science fiction experience the most pleasure by cognitively drawing the relation and imaginatively filling the gap between the imagined world and actual society (*Scraps of the Untainted Sky* 5).
4. While Luckhurst concludes that Robinson's style can be called proleptic Realism, "a modeling of the present day tilted five minutes into the future" (172), I would argue that rather than ascribing another genre label to Robinson's very generically diverse text, it is important to note that climate change provokes writers to move beyond one dominant genre or mode.
5. Moreover, as Cronon has pointed out, such an image has glossed over the fact that the American West, the epitome of wilderness, was indeed not a pristine uninhabited space but was turned into such a space by expelling Native Americans from that territory or killing them.
6. Since then, however, a lot of research has been done which counters this argument by showing that people living in rural areas are less likely to engage in preservation efforts—being far more dependent on the exploitation of the land as source of income—than city dwellers (see Bennett 40).
7. See for example *The Environmental Justice Reader* by Adamson, Evans, and Stein, or most recently Rob Nixon's *Slow Violence*. There

are also many insights in Lawrence Buell's elaborations on what he terms "second-wave ecocriticism," which "concentrated strongly, for example, on locating vestiges of nature within cities and/or exposing crimes of eco-injustice again society's marginal groups" (Buell, *The Future of Environmental Criticism* 24).

8. An OECD (Organization for Economic Co-Operation and Development) study from 2010 explains, "Cities consume a great majority—between 60–80%—of energy production worldwide and account for a roughly equivalent share of global CO_2 emissions" (17). At the same time, due to their proximity to coastal areas, many major cities are also extremely vulnerable to climatic changes.
9. In his insightful article on nature and the city, Lawrence Buell defines six key tropes—"the city/nature as binary; city as holistic macro-organism; city as fragmentary assemblage; as palimpsest (i.e., as stratification of layers over time); as network; and as apocalypse" (4)—that dominate urban discourse. While Buell rightly criticizes our dependence on thinking about urban nature in these terms, he, at the same time, feeds into such a hegemonic discourse construction by limiting his interpretation to exactly these "key tropes."
10. In his analysis of the *Science in the Capital* trilogy, Douglas Kilgore asserts, "Robinson avoids the all-too-common trope of a catastrophe that prompts pleasure in civilization's ruin and its reconstitution as socially purified survivalist utopia" (5).
11. Here I am borrowing a term from Marie Louise Pratt, who points out "contact zones" in the colonial context in order to describe spaces of interaction between formerly geographically and historically separated groups (Pratt 7–9).
12. Douglas Kilgore further elaborates this argument by pointing out that in Chase's speech, the United States figures as the "'child' of other nations [which] is a striking reversal that leverages America's 'nation of immigrants' character/legacy to authorize its actions in the name of all humankind" (1).
13. Similarly, Martyn Fogg, has described terraforming as the "the application of technology for the purpose of influencing the global properties of a planet" (in Pak 6). It is originally used in the context of an extraterrestrial planet that is made fully habitable for humans.
14. As an example for the latter, Heise draws on Emma Marris' *Rambunctious Garden: Saving Nature in a Post-Wild World*, which optimistically calls for embracing nature as a hybrid of wild as well as "managed" nature (17).

15. Chakrabarty problematizes the underlying universalism when using the term species for the discussion of the Anthropocene: "Why should one include the poor of the world—whose carbon footprint is small anyway—by use of such all-inclusive terms as *species* or *mankind* when the blame for the current crisis should be squarely laid at the door of the rich nations in the first place and of the richer classes in the poorer ones?" ("The Climate of History" 216). Yet since the Anthropocene threatens humanity as a collective, he also employs species—even though in terms of a negative universalism.
16. I employ the term "species" here to stress Robinson's attempt to portray humankind as profoundly embedded in the functioning of the planet.
17. Geoengineering, or climate engineering, is used to describe the "deliberate large-scale manipulation of the planetary environment to counteract anthropogenic climate change" (Royal Society 1). The Royal Society distinguishes between two categories of geoengineering: Carbon Dioxide Removal methods and Solar Radiation Management. However, both approaches can have unknown side-effects on regional weather patterns and do not work against the rising acidity in the oceans.
18. In her analysis of Robinson's trilogy, Johns-Putra elaborates that while Suvin sees the utopian as intrinsic to science fiction, the late twentieth and twenty-first century has seen a trend of dystopian science fiction. Robinson explicitly counters this trend by identifying himself as a utopian writer. He explains: "Utopia is a name for one course of history, a progressive course in which things become more just and sustainable over the generations. We're not there now, but depending on what we do, and what our descendants do, we could still be said to be living in a utopian history, as being on the path. I prefer to work as if that were the case. And it seems to me the great work continues" (qtd. in Johns-Putra "Ecocriticism" 751).

Representing the Underrepresented: Climate Justice and Future Responsibilities in Climate Change Fiction

Everyone had stopped joking about how "hot" it was. It didn't really matter how "hot" it was. The problem wasn't lack of water or an excess of heat, not really. The problem was that 4.4 million acre-feet of water were supposed to go down the river to California. There was water; they just couldn't touch it. (Bacigalupi 174)

Paolo Bacigalupi's short story "The Tamarisk Hunter" depicts how in a climatically changed future, stress on water as a scarce resource is exacerbated. The protagonist Lolo makes his living by ripping out tamarisks, which are growing along the shores of the Colorado River. Ever since "Big Daddy Drought" got hold of the country ten years earlier, access to water has become the main concern for the population living and depending on the river's water. In particular, after California had claimed the complete rights to the river, and, in several military interventions, blew up the water supply infrastructure of major cities like Phoenix, Lake Havasu City, Grand Junction, and Moab, survival has become extremely difficult for the few remaining people in the area. Lolo, however, believes that he has found "the secret to eternal life as a tamarisk hunter" (172) and thus to a continuous wage of "$2.88 a day, plus water bounty" (173), which is paid by the Interior Department: while cutting the tamarisks, which use up the river's water, he simultaneously plants new tamarisk roots in hidden parts of the corridor to be harvested in the future. Yet, the short story comes to a tragic close when two national guards pay Lolo a visit and inform

A. Mehnert, *Climate Change Fictions*,
DOI 10.1007/978-3-319-40337-3_7

him that the water bounty program has been cut. With most of the river "under control" and soon to be piped, the priority switches to seep and evaporation control of "the Straw," a huge water reservoir for California. Overnight, Lolo's "insurance plan" has lost its value, and deprived of any water rights, he, along with everybody else, is forced to relocate if he wants to survive.

Drawing on Bacigalupi's short story for the opening of this chapter serves to direct the focus of analysis to questions of climate justice, which tend to be dismissed in international political discussions dominated by economic and techno-managerial adaptation and mitigation measures. While global warming describes an *average* increase in temperature, causes and impacts of this transformation vary immensely on a national as well as international scale—an aspect that is seldom brought to the fore in graphic illustrations of rising greenhouse gas emissions. Unsustainable consumption and production patterns in industrialized countries contribute significantly to greenhouse gas emissions, whereas "the impacts will be most devastating to the majority of the people in the South, as well as the 'South' within the North" (International Climate Justice Network), who barely add to GHG emissions. The few international climate justice initiatives put in place so far focus on distributive justice between nations based on historical emission responsibilities.[1] Accordingly, conceptual tools such as "climate debt" were introduced to acknowledge past inequalities and to assert that industrialized countries are to be held liable for their excessive and exploitative practices.[2] The concept of climate debt grants that in relation to a fixed global carbon budget, poorer nations are "allowed" to emit greenhouse gases, while industrialized nations are in "debt" because of their surplus emissions. In addition, the Kyoto Protocol has aimed to account for developing countries' "right" to emissions—with its principle of "common but differentiated responsibilities" as well as the implementation of market-based mechanisms such as international emission trading or the Clean Development Mechanism. In economic terms and in the context of international politics, the idea of carbon debts serves as a helpful numerical instrument to introduce issues of global injustice into the climate change debate.[3] As Mike Ziser and Julie Sze explain, "Historical carbon debt is a powerful conceptual tool for placing GHG-emission restrictions and costs on those who have benefited the most in terms of development, ensuring that some measure of genuine social justice accompanies the global fight against GCC [Global Climate Change]" (404).

Emphasizing the international dimension of climate justice is an important way to counter protectionist tendencies and Malthusian arguments from what a growing number of advocates have called "lifeboat ethics." Originally introduced by American ecologist Garret Hardin as a reaction to population growth in the 1970s, his lifeboat ethics reject an equal sharing of resources on an international as well as intra-national level. Hardin not only opposed any development assistance to poorer countries, but also rigorously fought immigration. The sociologist Andrew Ross warns that today in the United States "the warmed over Social Darwinism preached by Hardin (a cofounder of FAIR [Federation for American Immigration Reform]) has found new acolytes in the fractious debates about global warming, fueling a backlash against the notion that rich nations owe a carbon debt to poor ones, let alone a humanitarian lifeline to swimmers trying to catch up" (*Bird on Fire* 242). James Lovelock, the originator of the Gaia hypothesis, has also expressed concerns about this trend. He wonders,

> [...] whom we can let aboard the lifeboats? And whom must we reject? There will be no ducking this question for before long there will be a great clamor from climate refugees seeking a safe haven in the few parts where the climate is tolerable and food is available. Make no mistake, the lifeboat simile is apt; the same problem has faced the shipwrecked: a lifeboat will sink or become impossible to sail if too laden. (248)

Literary critic Janet Fiskio points out that the "lifeboat" narrative has come to be regularly employed in climate change discourse. In her article about dominant narratives in global climate change discourse, she explains how Hardin's idea of the "tragedy of the commons"—"sharing a resource as a commons gives everyone the right to use it, but none the responsibility to care for it" (Fiskio, "Apocalypse" 18)—resonates in climate debates about the "problem" of climate refugees and Western countries' arguments about population control in developing countries. The global climate justice movement tries to counter these exclusionist tendencies, which restrict access to resources to a few privileged countries, and instead emphasizes international historical responsibilities.

Without wanting to diminish the importance of framing climate justice as a global issue, it should be noted that the "climate debt approach" constructs greenhouse gas emissions as something that can only be accounted for on a national level and thus has to be regulated by politicians. It forecloses a more differentiated discussion of the impacts of climate change because it elides intra-national class and race differences in its emphasis on national emission budgets. Framing climate justice as a singular discourse that can only be addressed by focusing on distributive justice among international actors not only neglects multiple local and national justice claims, but also perpetuates the idea that climate change is a problem that requires a techno-managerial and economic response. John Urry and Bronislaw Szerszynski further problematize this dominance of economics in climate debates:

> Because "economics" got in first, it has largely monopolized the way that the social is conceived in the discourses of climate change. It has led to a focus on human practices as individualistic, market-based, and calculative, and has thus helped to strengthen a tendency towards a certain set of responses to climate change, ones based on individual calculation, technology and the development of new markets. (3)[4]

Such responses have tended to neglect the social dimension of climate change. They do not account for local or national socio-cultural differences or community dynamics, which also greatly determine human behavior and responses to climate change. Furthermore, in foregrounding issues of distributional justice between nations, the United Nations Framework Convention on Climate Change (UNFCCC) and other international political players frame climate change justice as something ***added to***, rather than ***embedded within***, existing structural inequalities (Fisher 25). Environmental justice scholars, however, have identified the correlation between the distribution of wealth, power, and environmental hazards (Adamson, Evans, and Stein 5). Accordingly, climate change is likely to aggravate environmental injustices experienced by minority communities or by marginalized individuals within nation-state boundaries, and thus intensifies rather than simply adding to structural inequalities. Susannah Fisher concludes that the carbon debt approach functions as a supra-national mechanism, which presumes the continuation of intra-national environmental injustices. She explains, "The current mode of conceptualizing climate change adaptation within the UNFCCC for example conceives of climate justice as the resilience of existing social systems rather than the transformation to new more equal societies" (Fisher 25). In contrast, Fisher argues, a local and national focus can bring the multiple intra-national justice claims to the fore, and enrich as well as broaden the discussion of possible responses to the challenges that climate change poses. Ziser and Sze also affirm that a phenomenon of such complexity such as climate change needs to be discussed in frameworks "that account for individuals, communities, and cultural and racial contexts as much as for net emissions, capital flows, and global trade" (407). Hence, frameworks are needed that show how climate change affects those who are most vulnerable on a global as well as on a local scale.[5]

Consequently, planning for more sustainable futures requires an awareness of how environmental problems are also intricately intertwined with national issues of social injustice. In *Environmentalism in Popular Culture: Gender, Race, Sexuality, and the Politics of the Natural*, Noel Sturgeon insists, "We must think about social inequalities as much as we think about environmental problems, and we must understand their interrelations" (5). Environmental justice scholars have, most notably, exposed the intricate relationship of social and environmental problems. They have revealed how poor communities, women, and ethnic minorities are disproportionately exposed to contamination at workplaces or at home, how these groups are especially vulnerable to environmental disasters and often have only limited access to resources that would improve their living conditions. Environmental justice scholars have broadened the very meaning of "the environment" by challenging mainstream environmentalism for its overemphasis on the protection of "pristine" landscapes of wilderness and for failing to take into consideration the urban environments that poor and marginalized communities inhabit. Moreover, the environmental justice movement has aimed to counter tendencies that depict environmental problems as if they affect all humans equally, regardless of race, class, national, or gender differences (Ray 21–23). In exposing environmental injustices, the movement ultimately seeks to promote "environmental justice as the right of all people to share equally in the benefits bestowed by a healthy environment" (Adamson, Evans, and Stein 4).

Following these general ideas of environmental justice, the international climate justice grassroots movement outlined the *Bali Principles of Climate Justice*. While the principles focus on the global dimension of climate change and, similar to the UNFCCC, call for international emissions trading as a solution, "the core political and philosophical focus of this document is on the 'most vulnerable' people and places" (Sze, "The Question of Environmental Justice" 6). The Bali document emphasizes that the adverse effects of climate change especially threaten marginalized groups—women, the young and old, poor and indigenous people, and ethnic minorities—and calls for responses to climate change that address these structural inequalities. Yet, the principles provide no insights into what these vulnerabilities mean in particular people's lives or what specific justice claims they entail. This is, of course, due to the nature of such kind of a text. It is, however, also a result of the peculiar challenge that global warming poses for communication due to the divergence between its causes and effects—its spatially and temporally distant ramifications. Not only is it difficult to imagine how climate change affects societies and individuals across time and space; it is also difficult to cognitively and emotionally relate to these events. As communication scholar Susanne Moser has explained, "How individuals see themselves, their roles and actions, their rights and responsibilities fitting into the larger social and ecological whole are some of the 'moral uncertainties' that can perplex people" (36). More compelling narratives, rather than large-scale emission scenarios or declarations of principles, might thus be needed to help people navigate through a complex system of future responsibilities. One medium through which these multiple justice claims can be communicated and portrayed in depth is literature.

In their detailed depiction of personal experiences and community interactions in a climatically changed future, cultural texts contribute to an understanding of the socio-political and ethical ramifications of climate change. Drawing on Edgar Platen's and Mathias Mayer's idea of "ethical text cultures" Hubert Zapf elaborates on the power of fiction to provide the "specific means of communicating ethical issues in such a way that it simultaneously resists conventional moralizing" ("New Directions" 5). Climate change fiction may provide a nuanced way—going beyond the conventional (and often apocalyptic) good versus bad moral narrative—of depicting the ethical dimensions of a particular problem that society faces. Furthermore, they are particularly adept at generating emotion and empathy.

By demonstrating how climate change affects characters' social and personal interactions, and their efforts to emotionally and physically cope with disaster and loss, fiction can serve as a way for readers and viewers to empathize with people across time, and thus with future generations, as well as with people in different social, economic, and ethnic contexts. Affirming the importance of empathy for social change, philosopher Roman Krznaric posits that empathy is also crucial for "closing the gap between knowledge and action, and for tackling the climate crisis" (155). According to Krznaric, literature and film are particularly good as generating shared emotional responses and thus form part of a body of educational measures that may help to provoke action on climate change. Even though it may be difficult to verify Krznaric's claim that fiction generates enough empathy to cause behavioral changes, his argument is, nevertheless, of value because it emphasizes the importance of empathy in times of crisis.[6] As climate change is expected to exacerbate resource shortages, it may also seriously challenge social cohesion (Archer and Rahmstorf 229). Krznaric thus correctly stresses: "Empathetic recognition of other people's humanness is the most basic necessity for preventing the worst forms of cruelty and violence" (169). Other highly influential contemporary philosophers, such as Martha Nussbaum, have also stressed the importance of literature and the arts not only in providing resources for our own "incomplete sense of life" (*Love's Knowledge* 28), but also in generating compassion for other human beings.[7] Importantly, the non-human realm forms part of Nussbaum's "sympathetic imagining." She explicitly points to the importance of being confronted with different imaginings and worldviews, and challenged by the perspectives of silenced others. Climate change fiction renders these different perspectives visible while simultaneously promoting empathetic recognition.

While it is difficult for ecocritics, who do not engage in cognitive psychology or reader response theory, to account for the ways that literature generates emotions and compassion, an ecocritical analysis of climate change fiction can illustrate how issues of environmental injustice are framed and, ultimately, what particular justice claims are articulated.[8] Cultural productions such as films or literature can then serve as key sites that contest universalizing GHG narratives because they provide "insider perspectives" (Adamson, Evans, and Stein 9) on the struggle for climate justice and reveal what otherwise remains hidden in emission graphs—that is, the intra-national, social, and ultimately personal dimensions of environmental injustice. Rob Nixon also elaborates on the important role writers and filmmakers can play in conveying the "lived sense of a corroded environment" that marginalized communities inhabit in order "to counter the layered invisibility that results from insidious threats, from temporal protractedness, and from the fact that the afflicted are people whose quality of life—and often whose very existence—is of indifferent interest to the corporate media" (16). In order to illustrate how climate change fiction illuminates the manifold facets of climate injustice and calls attention to the close intertwining of environmental problems with issues of class, race, gender, and ability discussed in this chapter, I do not limit my analyses in this chapter to novels but draw on a range of cultural texts: Bacigalupi's short story, Dana Stein's novel *Fire in the Wind,* Octavia Butler's novel *Parable of the Sower*, and the film *Beasts of the Southern Wild*. As Phoebe Godfrey explains, "These inseparable social categories not only shape how we see and understand ourselves/other humans and the social and natural worlds, including other living beings, but also form the basis of how we see and understand the causes and current/future consequences of GCC" (5). It is thus imperative that they are considered as part of the discourse on climate change.

The cultural texts named above complicate the idea of one collective climate change narrative and instead remind readers of the disproportionate effects of the transformations that global warming will bring about. In the rest of this chapter, I analyze how these works counter this inequality by drawing on three dimensions of justice—distribution, recognition, and participation—and posit them as crucial in planning for more just climate change futures. For this purpose, I rely on political scientist David Schlosberg's discussion of environmental justice. In *Defining Environmental Justice*, Schlosberg critically notes that while most environmental justice debates have focused on distributive justice, an exclusive focus on maldistribution fails to address the underlying processes that lead to distributional injustice in the first place. He points to Iris Young's *Justice and the Politics of Difference* and Nancy Fraser's numerous publications that have also defined individual and social recognition as key elements of attaining justice (4) and insists that it is important to examine the context of oppression, and thus the "structural nature of the construction of subordinate and disrespected identities and communities" (15). Schlosberg stresses that the concept of recognition is not a substitution for distribution as an essential component of justice, but rather sees them as complementary. Like Young, he also views a third component—participation—as crucial to justice. He summarizes, "Both distributive inequity and misrecognition hamper real participation in political and cultural institutions ... In this case, improved participatory mechanisms can help meliorate both other forms of injustice; but those forms of injustice must be addressed in order to improve participation" (28).

Taking these three dimensions as a starting point for discussion is helpful in understanding how fiction participates in the communication of climate change as well as more clearly articulating what specific climate justice claims it makes. While the three dimensions of environmental justice—distribution, participation, and recognition—guide my textual analysis, it will become clear that no single text engages in solely one dimension.

Often all three dimensions are intricately intertwined and communally play a role in the texts, but as I try to show in my close readings, the dimensions are weighted differently. In all of the future scenarios depicted, current inequalities will continue and are aggravated. However, the texts provide different responses to this climate injustice. Works such as Butler's *Parable of the Sower*, for example, postulate that adaptation to climate change can serve as an opportunity for societal transformation. Her book thereby allows for a re-examination of the term resilience, "the ability of a system to absorb disturbance and still retain its basic function and structure" (Walker and Salt 1).[9] While the term has a rather positive connotation in ecocritical discourse and is used to give credit to forms of community resistance (Murphy "Community Resilience"), it has to be critically noted that even though it describes forms of scattered resistance, resilience assumes the continuation and maintenance of the same fixed societal structures. Read as such, and in contrast to Butler's novel, most of the works discussed in this chapter depict that even though adaptation to climate change might involve resilient developments, these will not be successful without a wider social transformation and a greater emphasis on climate justice.

Climate Resilience and the Creation of Sacrifice Zones

This section aims to illustrate how in climate change fictions such as Bacigalupi's "The Tamarisk Hunter" (introduced at the beginning of the chapter) and Dana Stein's *Fire in the Wind*, climate change aggravates processes of maldistribution. As a result of this distributive injustice, sacrifice zones emerge—sites that are deemed dispensable by people in power because they tend to be inhabited by poor and powerless people. In "The Tamarisk Hunter," a long period of drought has led to regional conflicts over the Colorado River and ultimately to one state's total control over it. Owning access to water, the single most important resource, California controls the livelihood of all the other out-of-staters who depend on the river and, as the story unfolds, it becomes clear that this resource is not fairly distributed. While the river once connected different regions, states, and people, and served as a symbol for mobility and border crossing, it has now been turned into the prime site of contestation, epitomizing exclusion and precariousness. The story here translates the actual struggle over water, which so far most prominently has taken place along the US-Mexican border, into an intra-national conflict and exemplifies how "water symbolizes the contested politics and the geographic and cultural spaces between [...] communities that hold unequal power" (Sze, "Boundaries of Violence" 476). The protagonist, Lolo, remembers that some communities tried to fight for their right to the river, but quickly adds that their efforts were doomed to fail against an overpowering military presence in the region. In a climatically changed future, the ownership of water becomes not only a tool of domination, but actually bestows its owners with power over and life and death. In this context of extreme distributive injustice, any form of resilience is made impossible, as epitomized by the failure of Lolo's attempts at resistance.

Moreover, the short story reveals how this injustice is the result of strategic planning by people in power. As a colleague explains to Lolo, "They figured out that they had to do it slow. They worked out some kind of formula: how many cities, how many people they could evaporate at a time without making too much unrest" (Bacigalupi 179). Ultimately, officials in the Californian government as well as the Interior Department decided which regions and people to sacrifice and which ones were worth preserving. "The Tamarisk Hunter" here illustrates that climate change does not pose a problem for the privileged few who live well-protected inside California's state boundaries. However, for the poor communities throughout the region that are deprived of their livelihoods, climate change poses serious challenges for survival—especially since California has begun prohibiting immigration and other states in the north are following its example. This leaves communities along the river not only without water, but also without any hope of finding a place where they can relocate. Furthermore, in depicting how this distributional injustice is strategically planned, Bacigalupi exposes how certain communities are deliberately targeted and deemed "dispensable citizens."[10] Referring to Rob Nixon's use of Annu Jalais' term "dispensable citizen" is helpful in this context, because it describes the hegemonic intentionality in the process of depriving people of a healthy environment and transforming it into toxic wastes sites or places of resource exploitation (Nixon 17).

Bacigalupi's dystopian story postulates a scenario of a future world in which differences—regional and economic—are aggravated as access to water becomes the primary concern and the Colorado River is transformed into an exhaustible resource to be exploited solely by the state of California at the expense of all living beings—human and nonhuman—depending on the course of its flow.[11] Given the fact that the

Colorado River is already not meeting the water demands of the growing populations along its shores, and that water assessment experts point out that the flow of the river will decrease by 5 to 20 percent over the next 40 years, Bacigalupi's story is of particular relevance and resonates actual developments.[12] Illustrating how the Colorado River, an essential artery for 40 million people living in the Southwest, ceases to be lifeline to all, "The Tamarisk Hunter" calls attention to the ethical ramifications of climate change and stresses the importance of taking issues of distributional justice, in this case exemplified by the struggle over scarce resources, into consideration. While resonating with Benjamin Franklin's famous dictum "When the well is dry, we learn the worth of water," the story also counters the delayed apprehension of the effects of climate change by imaginatively making this future accessible and bringing the injustice into focus. It thereby reminds readers that thinking about the future means thinking about fair distribution, especially with regard to those who are most vulnerable to global warming.

Similarly, Dana Stein's novel depicts how sacrifice zones emerge in the wake of climate change. The novel thus postulates that adaptation to climate change will lead to growing environmental injustice. In his book, Stein introduces a diverse set of characters in order to provide multiple perspectives on how climate change has altered the United States in 2036. The chapters shift between Harry Harper, a farmer in southeastern Iowa; Michael Baines, a member of the National Security Council, and his family; and finally, Margolis, a university professor, climate change expert, and environmentalist. This juxtaposition of a climate expert, a politician, and a farmer serves to depict climate change in its complexity and to illustrate in detail the effects of global warming on different groups in society. In doing so, the novel locates different sites of conflict and exclusion. While the narration is rooted in a middle/upper class perspective, it nevertheless highlights issues of environmental injustice based on class differences. More importantly, it reveals how climate change increasingly puts pressure on social cohesion. Stein's novel here not only draws on issues of distributive justice, but also points to participation in the democratic process as essential for a just society.

The chapters that trace the story of farmer Harry Harper illuminate how global warming leads to internal migration as places are stripped of their life-sustaining features. Once a fertile landscape, Harry's farm and fields, "which had been passed down through eight generations of Harpers, [were now] withering away" (14). While farmers in the region were initially confronted with heavy rainstorms, which still allowed them to cultivate their fields after they had implemented several drainage systems, the subsequent long-lasting drought soon forced them to either sell their land or to persevere while depending on money from the National Global Warming Relief Agency and hoping for an improvement in the situation. Yet, as the situation worsened and they were deprived not only of their livelihoods but also of their very homes, for many farmers there only remained one solution: suicide. Thus, when Harry receives a government letter stating that relief will no longer be paid, he knows that his only chance for survival is to abandon the place that has been his home for so long. In the hope of finding good land to farm, he heads west. On his way, Harry explains to another farmer and fellow traveler that he is no longer in contact with his son. He admits that for too long he was in denial that the "leaders intentionally got [them] into this mess" (Stein 57), and often argued with his son who already early on accused the government of this "betrayal, especially of his generation" (Stein 56). Even though the government had known about the impact of global warming, it did not act accordingly but instead accepted that some landscapes and the people inhabiting them had to be sacrificed.

Stein draws here on the transformation of the agricultural sector in the United States and depicts how farmers are dramatically affected by climate change. Once part of the nation's "breadbasket," providing many jobs and contributing to the nation's wealth, Iowa is now part of a Midwestern wasteland. Furthermore, in focusing on how "farming the land," long postulated as a core US value, is no longer possible, and on how even families as "patriotic as you can be" (Stein 57) begin to question their country because they have been deemed "dispensable" by their government, *Fire in the Wind* reveals how, along with fair distribution, "procedural equity" (Bullard)—the opportunity to participate in democratic decision-making—is also an important aspect of justice.[13] Schlosberg also affirms that "fair and equitable institutional processes of a state" (25), which allow for individuals' participation in determining the course of a community's common activities, are crucial to ensure social justice. In *Fire in the Wind*, democratic decision-making structures are increasingly rendered obsolete. Bringing the correlation between distributive injustice and procedural inequity to the fore, the novel shows how farmers in the area, who had been influential and respected based on the former wealth of their lands, had been driven toward obliteration as the value of their soil plummeted. Similar to Bacigalupi's short story, Stein's novel illustrates that sacrifice zones are created in accordance with levels of income and access to resources.

In the novel, other parts of the country besides the Midwest are also severely altered by climate change. When Harry finally reunites with his son, they both decide to move to California. However, Stein defamiliarizes the motif of moving to the promised lands of the West. California has been significantly transformed by wildfires and turned into the place most in need of firefighters like Harry's son. Yet, Harry believes that his own expertise as a farmer might also be appreciated in order to help all the other displaced people "who, like him, had their world turn upside down" (147) and to teach them how to farm their own "victory gardens." This reference to the victory gardens, which were promoted during World War I and II as way to help with the military effort, postulates climate change as a war that is fought at the home front.[14] However, as the novel shows, this time it requires a community effort that is not part of a campaign led by the national government, but rather one that works in defiance to the neglect and the creation of sacrifice zones by the national government. The gardens serve as purpose-giving spaces, which allow for self-fulfillment, societal participation, and provide hope for future generations.

This closer examination of how and in what places the United States is affected by climate change serves to illustrate the importance of space, or the lack thereof, in making justice claims. Following the story about NSC advisor Michael Baines in *Fire in the Wind*, the reader finds out that Michael's father, similar to Harry Harper, was deprived of his livelihood. Working as a fisherman of crabs and oysters in the Chesapeake Bay, he was forced to move when sea levels rose and increasingly destructive hurricanes hit the region. Baines recounts that while the government had agreed to build several sea walls, it "tended to be the more upscale communities" that got them. Poorer communities, however, lost not only their livelihoods but also their very homes, and ended up "living out of a trailer, or even pitching a tent, in one of the large resettlement areas that the Department of Homeland Security had set up outside cities" (Stein 19). "The displaced," as this group of people comes to be commonly called by the rest of society, were thus intentionally pushed into invisibility at the periphery. Stein's novel here exemplifies the role of space in the struggle for environmental justice and shows that in the aftermath of disaster, spatial discrimination continues. The people whose lands have already been deemed dispensable are relocated to areas that are equally worthless. Environmental burdens are thus purposely placed on disempowered and dispossessed communities. Furthermore, the novel illustrates that as low-income people and people of color are forced to move to designated areas outside of the city, they are not only denied access to the city, but also to infrastructural improvements in their own neighborhoods. When Michael visits one of the displacement camps outside of Washington D.C., the cab driver comments on the social division of the city when he notices "I've never had someone from this part of town ask for a ride to the Anacostia" (119). Upon his arrival, Michael, who had never been to the area before, is surprised at the horrible living conditions and compares it to "the type of development he'd seen a lot of in South America's poorer cities" (120). His visit is interrupted by a flash flood, which further illustrates how injustice is inscribed in the urban planning process. Since the displacement camp was built in a low-lying area, it is hit hard by the flooding, which occurs now with greater frequency. Michael barely manages to save some of the community's most vulnerable inhabitants—a couple of children and an old man. Yet while some communities are twice ostracized, displaced by multiple climate events, these dramatic circumstances remain unknown to the rest of the population. Because of the lack of necessary infrastructure, these individuals are structurally barred from making their claims heard among people living in cities. The ways in which climate injustice works are thus profoundly linked to asymmetries of spatial power.

Another passage in the novel further underscores that environmental injustice is a result of a systematic structuring of disparity. While passing through the outer neighborhoods of Baltimore, Michael, torn between the obligations of his job and his own moral conscience, notices, "Baltimore was home to some of America's finest hospitals and universities, but prosperity's advance remained stalled outside most of the city's older, low-income neighborhoods" (Stein 82). Pointing to the distributional as well as participatory injustice, this passage resonates with environmental justice research, which has revealed the correlation between political disempowerment and exposure to environmental threats. As sociologist Daniel Faber explains:

> The less political power a community of people possesses; the fewer resources a community has to defend itself with; the lower the level of community awareness and mobilization against potential ecological threats; the more likely they are to experience arduous environmental and human health problems at the hands of capital and the state. (4)

Stein's description of Baltimore exemplifies this spiral of injustice, while also alluding to the importance of access to education. The problem here is not solely one of unjust distribution, but again of institutionalized exclusion in which space plays an important role. So *Fire in the Wind* points to how climate change aggravates social disparities, as the government declares "sacrifice zones," which undermines participation in decision-making processes.[15] The motif of the wall brings this most notably to the fore. The construction of higher dikes in affluent areas, the wall that encircles Baltimore Harbor, and the walls that "had become the new landmarks of the southern tip of Manhattan" (41) all serve, on the one hand, as a physical barrier to rising sea level. On the other hand, they symbolize denial and exclusion. As climate change is banned to the "other side of the wall," it is made invisible. Moreover, the wall exemplifies that in times of climate crisis, only a certain part of the population is protected. Those, who remain outside the walls, the marginalized communities, are also the ones who actually "experience" climate change the most. According to Rob Nixon, "walls concretize a short-term psychology of denial: the delusion that we can survive long term in a world whose resources are increasingly unshared. The wall, read in terms of neoliberalism and environmental slow violence, materializes temporal as well as spatial denial through a literal concretizing of out of sight out of mind" (Nixon 20). In *Fire in the Wind* walls "concretize" climate change out of mind, and, at the same time, serve to exclude low-income communities from society.

By bringing the contestation over space to the fore, Stein's novel emphasizes that participation in the greater community is key to climate justice. Even though walls and sacrifice areas are created as a means to further exclude "the displaced," there are several people—Michael's father among them—who try to reclaim space. In occupying the National Mall, they counter efforts to render them (and thus the effects of climate change) invisible. At the same time, the group of protesters insists on their right to partake in the democratic decision-making process and thus not be excluded from decisions about relocation or about climate change action more generally. However, as the novel continues, the protest of "the displaced," as well as of several university students, and various environmentalist groups, becomes increasingly difficult and dangerous in a country that prevents political participation and closes down spaces for demonstrations. Michael's job transfer from international UN advisor on environmental refugees to advisor for domestic climate change developments in the National Security Council already foreshadows the political trend toward "securing" internal affairs. Even though Michael begins to sympathize with the environmentalists over the course of the novel as he becomes increasingly aware of the injustice induced by the government he is working for, and ultimately even reconciles with his father, the novel's dystopian ending forecloses any hope of restoring justice. As tanks roll into Washington D.C. to overthrow a democratically elected eco-friendly president, democratic participation is made impossible by military intervention and distributional injustice is likely to proceed.

Similarly to Bacigalupi's short story, *Fire in the Wind* leaves no hope for any successful resilient action, let alone any transformative adaptation to climate change. Instead, global warming dramatically aggravates social disparities. Both works of fiction point to unjust practices that stem from increased stress on scarce resources, such as limited accessibility to certain places and severe restrictions to the democratic decision-making process. In contrast to these narratives, the two climate change fictions I analyze in the following section show that adaptation to climate change can also involve a process of empowerment and possibly even societal transformation.

From a "Politics of Disposability" to a Climate of Change

As the climate change fictions discussed in this chapter demonstrate, adaptation to climate change ultimately poses questions about who is protected and what gets saved. In response to climatic disasters, adaptive measures often exacerbate social disparities. A recent example in the history of natural catastrophes illustrates the interplay between social asymmetries and environmental hazards: Hurricane Katrina and the set of events in its aftermath very clearly revealed how the underlying social and racial stratifications of the affected city greatly influenced the outcome of the catastrophe.[16] As cultural critic Henry Giroux explains, "what first appeared to be a natural catastrophe soon degenerated into a social debacle" (173), especially when the media began reporting on the victims of the hurricane.[17] These images not only powerfully illustrated the plight of the people most affected by the storm by showing images of people stranded on rooftops or, worse, of corpses floating in the water, but also brought to the fore that most of the victims were African-American and poor. In the weeks after the flooding, reports increasingly focused on the inhumane conditions that people had to endure in the "refugee camps" (Brooks) in the city's convention center and its sports stadium, The Superdome.[18] The reverberations of the hurricane revealed something long ignored in the United States: that the problem of the "color line" still existed. More vigorously even, Giroux argues,

> Something more systemic and deep rooted was revealed in the wake of Katrina—namely, that the state no longer provided for the poor, sick, elderly, and homeless. Instead, it had been transformed into a punishing institution intent on dismantling the welfare state and treating the homeless, unemployed, illiterate, and disabled as dispensable populations to be managed, criminalized, and made to disappear into prisons, ghettos, and the black hole of despair. (175)

Giroux here criticizes the failure of the state, which instead of providing relief for the most vulnerable seemed indifferent to the suffering and to racial injustice. For Giroux, this dynamic reflects a political phenomenon he calls the "new biopolitics of disposability" (175).[19] Drawing on Angela Davis, who describes how the normalization of torture practices in domestic prisons in the United States has turned these places into sites "where democracy has lost its claims" (124), Giroux cautions that the events in New Orleans showed similarly repressive features. Furthermore, the initial governmental delay in providing relief actions, the exercising of military power, and the criminalization of African-Americans in particular was not only an example of a growing authoritarianism, but also demonstrated that the state privileged some lives over others.[20] The attempt to thereby render the poor as well as people of color invisible, to excommunicate them from the sphere of public concern, according to Giroux, reflects the country's newly implemented politics of disposability (175). Only three years after Hurricane Katrina the underlying social inequalities were no longer an issue for the media, nor of greater concern for the rest of the United States (Möllers 332). This endless position of neglect combined with the "atrophy of the public imagination," according to Giroux, leads to political disempowerment and forms a central element of the biopolitics of disposability (190). Underscoring the power of fiction, he thus calls for a greater critical engagement with these issues in the public realm, in order to open up spaces for social empowerment and critical discussion.

My brief discussion of the events in the aftermath of Hurricane Katrina serves as a transition to my analysis of the 2012 released and Academy Award–nominated film, *Beasts of the Southern Wild*. Benh Zeitlin's movie echoes the environmental injustice aggravated by Hurricane Katrina. The story focuses on a poor and multiethnic community, and in particular on two of its inhabitants—Hushpuppy and her father Wink—who try to survive in a small bayou settlement in Louisiana that comes to be inundated after a heavy storm passes through. While the film was praised for its original screenplay, its use of magical realism, and the brilliant acting of its six-year-old protagonist, Quvenzhané Wallis, it also received harsh criticism from scholars such as bell hooks for its reaffirmation of gender and racial stereotypes and its "pornography of violence" ("No Love in the Wild"). hooks argues that the movie's over-engagement with myths and magical realism masks its racist tendencies and plays down the traumatic experience of a child becoming an orphan. While hooks's as well as Agnes Woolley's critique of the film as reproducing hateful stereotypes about black masculinity may be partially justified, it would be wrong to readily dismiss the film as just that.[21]

Instead, *Beasts of the Southern Wild* is of importance for ecocritics and especially for critics interested in the representation of global warming because of its innovative aesthetics and its commitment to environmental justice. Ecocritic Patricia Yaeger, for example, not only found the movie thought-provoking for its portrayal of "luminous trash," but also because she sees in its mythopoeia a possibility to take "ownership for all the unintended consequences we have to live with … and a pedagogy out of our cosmic impasse." According to Yaeger, the film's potential lays in its Anthropocene images and in exposing the paradoxes of this "carbon-drunk age." The opening scenes of the movie, for example, juxtapose a close-up of a polluting factory, possibly an oil refinery, with aerial shots of what Wink claims to be "the prettiest place on earth" (00:04:33). Challenging viewer's expectations, the camera zooms in on this place and on what at first seems like scattered pile of trash from afar. It thereby juxtaposes the otherwise pristine image of the bayou landscape, with what actually turns out to be Hushpuppy's home: a community built out of shacks called the Bathtub. The subsequent carnivalesque imagery of the community festival again leaves viewers troubled. Despite the presence of beautiful and colorful camera play, close-ups of improvised costumes made out of garbage, of dirty places and bodies, of drunken parents, and abandoned babies remind spectators of the poverty in the community. Finally, landscape shots of the wastelands created by the storm, alternated with scenes depicting destroyed houses and dead animal bodies, forcefully bring the vulnerability of the community to the fore. The film provides more examples than the ones given here, all of which serve to counter Giroux's proclaimed atrophy of the public imagination. Yaeger affirms that the movie "creates a zone of history-making for Katrina's disposable bodies." Furthermore, she argues, "the film's rags and wastelands—its killing fields—become powerful emblems of the Southland's (and our nation's) commitment to toxic inequality." The movie thus forms an aesthetically innovative and critical response to the events of Hurricane Katrina.

Important for my discussion and reason for including this film in the chapter is its engagement with the issue of climate change. *Beasts of the Southern Wild* employs climate change not solely as a setting, but uses it to deal with questions of environmental justice and future responsibilities. Already in the beginning of the film, the viewer is made aware of the exclusionary practices that characterize Hushpuppy's life in the bayou settlement. Floating with her father on their raft made out of old car parts and other recycled materials, they pass the wall, which Hushpuppy knows was built by the people up "above the levee, on the dry side" (00:04:52) to cut them off. Again, the wall symbolizes not only the denial of the materiality of climate change, but also reveals the socio-spatial asymmetries of the Louisiana landscape. While Hushpuppy and her father barely survive on the wet side of the wall, living off the trash produced on the dry side, and are more dangerously exposed to the effects of climate change because of the levee, the people on the other side are safe. The wide shot about four minutes into the movie powerfully visualizes the danger: the numerous chimneys emitting gray clouds full of CO_2 tower over Hushpuppy and Wink, who seem vulnerable and small on their raft on the other side of the wall surrounded by dirty water. Here, *Beasts of the Southern Wild* exposes the distributional injustice that already existed before the climate disaster. This social and racial disparity is intensified when the storm hits land.

The movie leaves no doubt that the hurricane, which devastates the Bathtub, is related to global warming. As Hushpuppy's teacher, Mrs. Bathsheeba, explains to the children in her floating school, "Fabric of the universe is coming unraveled. Ice caps gonna melt. Water is gonna rise. And everything south of the levee [thus also where the Bathtub is located] is going under" (00:09:28–30). After her teacher's warning, Hushpuppy looks at the yellowed image of the South Pole pinned to the wall of her classroom and her voiceover begins to recount the tale of the mythic aurochs, a prehistoric giant pig-like beast, which was captured in the ice during the Ice Age and which now might be released from its glacial imprisonment.[22]

The aurochs function as an intelligent stylistic device for the representation of global climate change. Drawing on their story, Hushpuppy is not only able to connect the events happening in the Bathtub to the occurrences at a completely different place—the Poles. More importantly, the beasts symbolize the long-term temporal dimension of climate change. While the storm that passes through Louisiana is a singular and momentary event, the story of the aurochs stresses its long-term implications. Furthermore, it reminds viewers of the profound impact of human actions that had the power to disrupt geological timescales. As aurochs, figures of the pre-historic past, come to haunt our present and future, the film challenges short-term present-day thinking and places the storm and its aftermath in its geohistorical continuity. Pointing to challenges that climate change poses for representation, Dipesh Chakrabarty has argued, "Our usual historical practices for visualizing times, past and future, times inaccessible to us personally—the exercise of historical understanding—are thrown into a deep contradiction and confusion" ("The Climate of History" 198). This confusion is the result of the current sense of crisis "that disconnects the future from the past by putting such a future beyond the grasp of historical sensibility" (197). *Beasts of the Southern Wild* restores this historical sensibility and enables a reflection on human involvement in planetary processes through its employment of magical realism. Without having to revert to blockbuster effects of huge tidal waves or the coming of the next Ice Age, the film maintains its "realist" and sometimes even documentary-like camera style, but introduces magical elements that highlight the threat that climate change poses. The aurochs thereby serve to innovatively scale up the human imagination.

Moreover, the aurochs help to articulate social and political critiques. I thus disagree with Agnes Woolley who argues that "the magical-realist mode of *Beasts* erases questions of racial, gender and environmental justice through abstraction and mythologisation" ("The Politics of Myth Making"). On the contrary, the aurochs epitomize the social disparity caused by climate change. As Hushpuppy's voiceover informs the spectator, they are "strong animals [who] know when your heart is weak" (00:33:38), and show no mercy toward those who are vulnerable. As climate change causes melting at the Poles, the aurochs are released from the ice and along with the water masses, and they ravage their way across the country on their destructive path toward the Bathtub. Metaphorically, the aurochs symbolize climate change, and give this invisible and elusive phenomenon a shape and body. Furthermore, highlighting that global warming dramatically affects the poor and minorities, it is no wonder that these creatures seek out Hushpuppy's community. The danger of the approaching aurochs aggravates the already precarious situation, since the community is facing the complete destruction of their homes and the loss of their very livelihoods after the hurricane has passed through and the high level of salt in the water has killed plants and animals alike. Wink's mysterious medical heart condition has also worsened. The aurochs, representing humankind's bad decisions in the past, are now haunting the present. Yet, epitomizing the injustice that accompanies climate change, these creatures do not seek out the people above the levee who contributed most to rising greenhouse gases, but first come to the poor and vulnerable communities.

However, while the community has already been forcefully displaced once before and relocated to the "Open Arms Processing Center for Disaster Relief and Response," Hushpuppy now stands her ground when faced by the aurochs. At the processing center, she had noticed how "It didn't look like a prison. It looked more like a fishtank without water" (01:02:23), and thus emphasized that because of this forceful displacement, she and members of her community had been deprived of their very livelihood. Furthermore, the shots of the center resonate with the pictures taken of the convention center after Hurricane Katrina, and visually remind viewers of these past injustices. In the movie, white doctors are looking after mostly black "displaced" people. The community's disempowerment is symbolically completed when Wink is narcotized and thus made defenseless. Hushpuppy watches as the nurse pushes her father around and stuffs pills into him. Seemingly detached from the events, Hushpuppy's voiceover describes, "When the animal gets sick here, they plug it into the wall" (01:05:12), describing the dehumanizing processes at the center. Here, the movie prominently brings an oppressive pattern of cultural domination and non-recognition to the fore.[23]

While distributional injustice is a visually dominant aspect in the movie, Zeitlin shows that lack of recognition also contributes to climate injustice. Recognition is understood psychologically as well as structurally. Drawing on Axel Honneth, Schlosberg clarifies that lack of recognition is injustice because "it impairs persons in their positive understanding of self—and understanding acquired by intersubjective means" (18). Fraser, however, has argued that misrecognition is foremost related to status and thus dependent on structural, social, and symbolic indicators (ibid.). Both dimensions of misrecognition, the psychological as well as the structurally constructed, play out in the processing center scene in *Beasts of the Southern Wild*. Not recognizing the Bathtub community's way of life, Hushpuppy, her father, and the others are not considered as valuable members of society and thus not included in the decision-making process. Instead, the disrespect toward the community is expressed in their forceful displacement and in the racial discrimination powerfully visualized in the scenes about the processing center. Wink's narcotization further exemplifies how self-determination—and, consequently, a positive understanding of self—is made impossible. In the end, Hushpuppy defies this social and racial subordination. She knows that this is not the way her father wants to die and so they escape from the center and return to the Bathtub. Having this experience still in mind, Hushpuppy thus turns to face the aurochs when they make it to the front porch of the last house standing in her community.

The encounter with the aurochs epitomizes the struggle for recognition and empowerment. Even though the huge beasts come threateningly close to Hushpuppy, she shows no fear but stands firm and protects her community. The beasts finally bend down in front of her and she observes, "You're my friend, kind of" (01:22.14). This scene marks a turning point: here, Hushpuppy rejects any further victimization that the arrival of the beasts—climate change—might instigate. Instead, defying further misrecognition and injustice, she emphasizes that her community is not the one that should be punished or bear the costs of the climate crisis. Her few words, ultimately, signal another important theme of the movie: the close entwining of humans and their non-human surroundings. "With this statement she acknowledges that nature is both friend and enemy" (hooks) and reaffirms a "cosmic consciousness" (ibid.) present throughout the entire film.[24] While humankind's complicity in the destruction of the planet is a given, the movie more prominently reveals the social injustice that goes along with such destruction.

The story leaves viewers uncertain about Hushpuppy's future, especially after her father's death. Similarly to *Fire in the Wind*, Zeitlin's movie depicts the failure of older generations to take responsibility for future generations. Yet, *Beasts* nevertheless ends on a hopeful and uplifting note because it puts faith in younger generations to reclaim their rights and spaces, to show resistance. Ultimately, it is Hushpuppy who leads the rest of the community, raising a flag and defying the surf crashing on the pier. At the film's close, Hushpuppy realizes,

> I'm a little piece in a big, big universe. And that makes things right. When I die, the scientists of the future they're gonna find it all. They gonna know once there was a hushpuppy and she lived with her daddy in the Bathtub. (01.05.42–05.43)

Beasts of the Southern Wild thus provides a space of recognition in the public sphere for the communities and individuals most vulnerable to climate change and postulates that this recognition is key for restoring justice. Furthermore, the narrative style of *Beasts* and its close resemblance to storytelling underscores that while Hushpuppy's story serves a didactic purpose, it also functions as a story of resistance that counters universalist discourses about climate change. The emphasis on recording Hushpuppy's story, through her self-portraits on cardboard so that she "ain't gonna be forgotten" (00:17:23) illustrates the importance of making individual climate change stories and climate justice claims heard. Fiction thereby critically reflects on consequences of distributional injustices and serves as a reminder of the people and places that suffer the most from the lack of recognition. While Zeitlin's movie remains ambivalent about the success of Hushpuppy's resistance, it nevertheless counters tendencies that aim to render marginalized communities invisible and calls for recognition of their experiences of climate injustice.

The utopian impulse of community empowerment in the aftermath of disaster is also an integral part of Octavia Butler's *Parable of the Sower*. While the first part of Butler's book portrays how global warming has dramatically exacerbated social disparities, to the extent that it has turned the United States into a place where democracy no longer has any claim, Butler's story transcends the apocalyptic "collapse of civilization" narrative and suggests that if racial, class, or gender differences are overcome, new communities can be formed. In fact, *Parable of the Sower* suggests that the environmental crisis can only be resolved if the social crisis based on exclusionary practices is also addressed. Butler ultimately envisions a sustainable future that can only be achieved through processes of social transformation.

Set in California in 2024, *Parable of the Sower* depicts how climate change and resource shortages have profoundly altered the US economic and political landscape: poverty, crime, and slavery prevail outside of a few gated and militarized communities. Multinational corporations have replaced government institutions in trying to maintain order—but at the price of repealing human rights (Stillman 15). These corporations offer safe cities in exchange for labor exploitation and debt slavery. Here, the story exposes the biopolitics of disposability, or what Mbembe calls "necropolitics," which he defines as "sovereignty resid[ing], to a large degree, in the power and the capacity to dictate who may live and who must die" (Mbembe 11). In *Parable of the Sower*, necropolitics function on the basis of racial and economic inequalities. While some of the richer communities threatened by rising sea levels receive support, others, such as the protagonist's community, are "too big, too poor, too black and too Hispanic to be of interest to anyone" (Butler 120). After years of drought and impoverishment due to maldistribution of resources, and in light of the growing violence in the country, many people migrate north toward Canada, which promises fertile soils as well as political and economic stability. At the onset of the story, the young African-American protagonist and homodiegetic narrator, Lauren Olamina, lives with her family in Robledo, a small gated community on the outskirts of Los Angeles. Even though Lauren's father and her stepmother both have jobs as teachers, the family is barely surviving, as Lauren's stepmother explains, "Lights, progress, growth, all those things we're too hot and too poor to bother with anymore" (5). Still, they have more than most people on the other side of the wall, the "street poor—squatters, winos, junkies, homeless people" (*Sower* 10) and so they try to protect their community as best as they can. However, over the course of the narration the situation worsens: Los Angeles is not the only city turned into a "carcass covered with too many maggots" (9) as private security armies are no longer able to protect communities and individuals from "thieves, rapists, and cannibals" (289) or worse from the "crazies" (110), a group of pyromaniac drug addicts, and Lauren realizes that "things are unraveling, disintegrating bit by bit" (123). While her father, who is also a Baptist priest and thus the community's spiritual leader, adheres to old routines and belief systems—for example, he is skeptical that God would have allowed something like global warming to happen—and still hopes that things will improve, fifteen-year-old Lauren is aware that this approach is a mistake. Describing the magnitude of the current transformation to her friend, "People have changed the climate of the world. Now they're waiting for the old days to come back" (57), Lauren knows that restoring the former order is impossible.

Indeed, the illusion of security that their community walls provided is shattered when intruders make it inside and rape and kill Robledo's inhabitants before almost completely burning down the entire community. This is when Lauren's coming-of-age story takes an important turn. With two other survivors, she embarks on a northbound journey on Highway 101, the "royal highway" of the past, which now is a "highway, a river of poor. A river flooding north" (223) and, for the first time in her life, begins to tell others about and teach them her religion "Earthseed."[25] Sylvia Mayer explains that Lauren's system of beliefs "consists of ontological and ethical ideas which center around the basic theological dogma 'God is change'" ("Genre and Environmentalism" 177). The Earthseed aphorisms scattered throughout the book thus emphasize flux and transformation, "interconnectedness and interdependence with the human and the non-human world" (Mayer, "Genre and Environmentalism" 186). By acknowledging a world of constant change, Earthseed goes beyond hierarchical or dualistic thinking and reasoning. And while change is the dominant force at play, Lauren, nevertheless, points to the possibility of shaping this transformation: "God *is* change, and in the end, God *does* prevail. But we have something to say about the whens and the whys of that end" (*Sower* 295). The direction and purpose of change can thus be influenced by humankind. Through Earthseed, Lauren provides a utopian vision to strive for and a meaning-giving purpose for the group traveling with her. At the same time, Earthseed heralds a new form of community interaction, one in which everyone is change and thus everyone can make a difference. Throughout the novel, Butler describes the grim situation Lauren and others face in the United States: constitutional laws were repealed in the 2020s, violence has destroyed most participatory democratic institutions, and voting has declined in importance among poor communities whose main concern is survival. It is in this context that Earthseed serves as an alternative model to the procedural injustice that has taken hold over the country. While the trip north also functions as a religious initiation for Lauren as well as her transformation from a poor, marginalized African-American girl to being a spiritual leader, she does not assume superiority over her fellow group members. Even though she invented the religion, she insists on the importance of group discussion and meetings in which decisions are made communally and the ideas of Earthseed are debated. Moreover, everybody is granted the right to voice his or her concern and to shape the course of direction of Earthseed.

In focusing on Lauren's journey and her teachings of the Earthseed religion, the book not only proposes a different understanding of the human-nature relationship, but more importantly reveals the community-building impulse based on group recognition and participation that may arise from disaster.[26] While violence toward minorities and a growing lawlessness in general has led to mistrust and self-reliance for survival, Lauren embraces the difference and diversity of her co-travelers. Even though multiethnic and mixed-race couples "catch hell" (171) out on the streets and weak travelers are the first ones to be attacked, Lauren is particularly drawn to these vulnerable groups. She thus invites a mixed-race couple, "a black man, a Hispanic-looking woman, and a baby who managed to look a little like both of them," an elderly African-American who will later become her husband, and a woman of Japanese and African-American descent and her daughter to join their group. Lauren is convinced that even in this world at the edge of social and ecological collapse, "friendship mattered" (182) and confides to her friend Harry, "It's not so bad if people stick together" (ibid.). Defying the radicalization of society, the growing individualism, racial segregation, and the fierce fight over scarce resources, the protagonist successfully builds a community that values sharing and trust and supersedes all ethnic, gender, capability, and age differences. Butler here explicitly perpetuates the idea that recognition and respect of the other is an important aspect of more just societies. Nobody can be excluded from participating in socio-political decision-making on the basis of gender, race, age, or ability. Convinced by Lauren's ideas of justice, many people start to follow her and ultimately found the first Earthseed settlement called Acorn. A passage from Earthseed's "bible," The Book of the Living, reflects best Lauren's commitment to diversity:

> Embrace diversity.
> Unite—
> Or be divided,
> robbed,
> ruled,
> killed
> By those who see you as prey.
> Embrace diversity
> Or be destroyed. (Butler 196)

Parable of the Sower reveals here that the ecological crisis is mainly also a socio-political crisis. While climate change is clearly the reason for water and resource scarcity, the book's emphasis is on critiquing the practices of exploitation and maldistribution that result from environmental degradation and that exacerbate social injustices. Yet, instead of remaining in the dystopian mode of presenting civilization's decline, Butler focuses on how in countering the politics of disposability, not only resilient but also larger "communities of change" can be formed.

She thereby postulates that an environmental crisis does not necessarily have to lead to exclusion and unjust practices, but that it can serve as an opportunity for societal transformation. Even though circumstances in *Parable of the Sower* are dramatic, this state of emergency seems to enable new ways of interacting with one another. Crisis does not stop people from making a communal effort but, on the contrary, it prompts them to try something radically new. Drawing on Rebecca Solnit's research about how "disaster destabilizes conventional ways of interacting, thereby allowing the formation of new social movements and collectives" (Fiskio, "Apocalypse" 23), Janet Fiskio shows how *Parable of the Sower* also offers a vision of "the collective" emerging in the wake of disaster. The suspension of the normal order, exemplified most notably in the downfall of Robledo's walls, forces Lauren to face the real world—the chaos "outside"—and, ultimately, empowers the protagonist to put her spiritual beliefs into practice and to create something new. As Solnit explains for post-disaster communities, "if paradise [...] arises in hell, it's because in the suspension of the usual order and the failure of most systems, we are free to live and act another way" (Solnit 7). The adaptation to a new order is, however, not a process without tension or problems. By depicting the initial internal conflicts in Lauren's group of travelers and their mistrust toward newcomers, Butler shows that community building is difficult because it requires the members' openness and willingness to see change, transformation, and difference positively. Even though members of Lauren's group have suffered violence, personal loss, and deprivation, Earthseed teaches them to trust and to "look out for one another" (Butler 223) despite ethnic, racial, gender, or age differences. Referring to Butler's depiction of a post-identity politics world, Peter Stillman explains,

> According to Earthseed, we are "potential" (*Talents* 361), with all the openness and possibility that word implies. Standard differentiating attributes—race, age, class, sex, religion, marital and familial status, sexual preference, personal history: Earthseed breaks these down, ignores them, reshapes them, and breaks them down again. (28)

Butler's utopian vision of more just climate change futures is based on the recognition and acceptance of these "potentials" of the other, because in accordance with Earthseed's emphasis on change, diversity can never amount to essential or insurmountable differences since identities are always performative, in flux, and constantly shaped anew.

In *Parable of the Sower*, emotion and empathy are important aspects in facilitating the creation of more just societies. Butler juxtaposes a system in which people show no compassion for human suffering with her protagonist's extreme sensitivity to the pain of others. Lauren was born with hyperempathy syndrome, an "organic delusional syndrome" (Butler 12) that makes her share the pain and pleasure of others (including animals), and which was caused by her mother's drug abuse when she was pregnant. Even though Lauren initially sees her "sharing" as a disability because it makes her more vulnerable, especially when confronted with the violence and pain outside the community walls, she later on considers hyperempathy syndrome as an asset that could help people become more conscious and pay particular attention to the vulnerability and suffering of others. Lauren's disability thereby comes to be reconfigured as a capability.[27] Musing about the senseless murdering, torture, and slave-like systems out on the streets, Lauren argues,

> If hyperempathy syndrome were a more common complaint, people couldn't do such things. They could kill if they had to, and bear the pain of it or be destroyed by it. But if everyone could feel everyone else's pain, who would torture? Who would cause anyone unnecessary pain? I've never thought of my problem as something that might do some good before, but the way things are, I think it would help. (Butler 115)

The protagonist realizes the ethical potential of hyperempathy syndrome. Sharing other people's emotions and feeling their pain may "serve the moral good" (Wanzo 74) to recognize the needs of the other. In fact, as critic Jerry Phillips states, "in a hyperempathetic world, the other would cease to exist as the ontological antithesis of the self, but would instead become a real aspect of oneself, insofar as one accepts oneself as a social being" (306). *Parable of the Sower* further stresses the importance of empathy in times of crisis by portraying how other "sharers" join Lauren's group and how they, in spite of their vulnerability, learn to trust one another even though they are complete strangers. Butler thus affirms that empathy is a fundamental aspect of a just society, which recognizes and values the difference of its members.

The employment of the motif of "hyperempathy syndrome" in Butler's speculative fiction is highly relevant for the discussion of climate justice. Even though references to climate change become less explicit over the course of the book as issues of social injustice are foregrounded, *Parable of the Sower's* focus on shared emotional responses allows for the problematizing of a particular challenge of climate justice. As discussed in the beginning of the chapter, the effects of global warming are most devastating for poor and marginalized communities. Scholars such as Krznaric conclude that empathy with those who are vulnerable is an important impetus for action against climate change. Butler's fiction also explores how global warming aggravates distributional injustice and how ethnic and economic minorities are further ostracized. At the same time, *Parable of the Sower* counters these exclusionary and unjust practices by highlighting the potential that a shared emotional response carries. By stressing the importance of empathically relating to the suffering of people in different social, economic, and ethnic contexts, the motif of the hyperempathy syndrome appeals to the moral responsibility of the reader in times of climate change. The awareness—in Butler's case epitomized in shared feelings—of the other's vulnerability thus calls for an ethical consciousness. William Edelglass's elaborations on Levinasian ethical thought are insightful in this context, as Edelglass posits that responsibility for the consequences of climate change emerges in the moment of becoming aware of the other's suffering. He further explains that "according to Levinas, it is a responsibility that arises precisely beyond any economic order, beyond history or cultural context" (221) and thus does not legitimize turning away from the suffering that is experienced by people across time, space, or socioeconomic context.[28] Since suffering can never be rationally justified, it morally obligates the person who is aware of the vulnerability of the other to act (ibid.). In *Parable of the Sower*, the hyperempathy syndrome denies even the possibility of ignoring the vulnerability of the other—at least if the suffering is visible—and symbolically reminds readers to recognize and to take into consideration those who are suffering from environmental injustices.

Octavia Butler's speculative fiction thus functions as a cultural space that prompts a reflection on what is unrepresented and neglected in current political discourses about climate change adaptation. Moreover, it counters those politics that aim to excommunicate the voices of marginalized communities. It accomplishes this by focusing on the groups that are vulnerable and thereby calling attention to the fact that they face disproportionate effects from global warming. The climate crisis is here foremost a social crisis, which can only be addressed by a reconfiguration of those underlying social systems that so far have served to promote injustice. Yet Butler's work is not only a social critique. By offering a utopian vision based on a communitarian and agrarian society, it explores alternatives to current environmental injustices. Moreover, Sylvia Mayer affirms that the title of Butler's work "places it in the tradition of another genre that is defined by its openly moral and didactic purposes" ("Genre and Environmentalism" 178): the parable. *Parable of the Sower* echoes the book's numerous references to the Bible and stresses the text's commitment to a moral claim. This moral claim is further underscored, Mayer adds, if one takes into consideration the book's parallels to the African-American slave narrative: the first-person narrative, the motif of the northbound journey, explicit references to slavery, as well as the manifold intertextual references to Bible passages place Butler's works in the tradition of the neo-slave narrative.[29] Rather than "estranging" the reader—as characteristic of speculative fiction—the book's allusion to the continuation of injustice in the course of history combined with its near-future and realist setting, emphasizes the story's similarity to the readers' world. Again referring to the didactic and moral aim of the text, Butler stresses here that "business as usual" cannot continue if one wants to avert the dystopian future the book presents. Instead, "an environmentally sustainable civilization must be based on the transformation of still dominant social power dynamics and their conceptual source of legitimization" (Mayer, "Genre and Environmentalism" 194). Distributional justice, recognition of difference, and the possibility of influencing the decisions that affect one's lives all are important aspects of the community that Lauren seeks to build and spread all over the country. Consequently, in *Parable of the Sower*, climate change adaption must thus go along with this kind of a societal transformation.

However, Butler's work is not a blueprint for utopia. Similarly to *Beasts of the Southern Wild*, *Parable of the Sower* ends on a hopeful note but does not provide closure, leaving readers uncertain about the destiny of Lauren's community. Even though the group around the protagonist is successful in founding Acorn, it is unclear if their enclave can survive in a country where climate change, anarchy, and violence are still dominating people's lives. Yet, it is especially this ambiguity that makes Butler's fiction so intriguing, because it insists on questioning what kind of a future a society wants to achieve for itself. Furthermore, Stillmann affirms, "Butler does not present a single utopian answer, but she does propose utopian possibilities" (32). These "possibilities," in contrast to generalizing scenarios commonly found in climate change discourse, acknowledge that global warming is likely to aggravate environmental injustices and social disparities. *Parable of the Sower* here proposes solutions that go beyond the techno-managerial fix, and instead points to the importance of addressing social and racial inequalities. In both *Parable of the Sower* and the sequel *Parable of the Talents*, Butler does not call for the establishment of some scattered, resilient communities, but instead perpetuates the idea of a large-scale transformation of society based on the recognition and respect of difference; a society in which all members have the possibility to equally participate in the decision-making process and in which resources are shared equally. Finally, Butler's is one of the few authors of climate change fiction who not only makes visible the plight of those most vulnerable to global warming, which forms an integral part of how we understand the future effects of climate change, but also envisions what possible adaptation plans could be derived from these conditions to counter climate injustice.

In articulating and critically discussing power relations in a climatically changed United States, the works discussed in this chapter draw attention to those issues that tend to be "excommunicated" from climate change debates. They thus foreground issues of justice, and engage with aspects of climate change that are difficult to depict in climate graphs and scenarios based on averages: they provide insights into the emotional and personal dimension of the disproportionate effects of the climate crisis, thereby calling into question discourses which perpetuate the idea that the human species is facing climate change collectively. Set in the United States, these texts pay particular attention to how intra-national injustices are exacerbated by climate change. While "The Tamarisk Hunter" and *Fire in the Wind* focus on issues of distributional injustice as a result of scarce resources, and Stein's work also addresses how participation in political decision-making becomes increasingly difficult, *Beasts of the Southern Wild* and *Parable of the Sower* foreground unjust practices based on lack of recognition and disrespect. Ultimately, these texts address all three dimensions of climate injustice—maldistribution of resources, limited participation in decision-making, and lack of recognition of others.

Countering any moral uncertainties that climate change might prompt, the narratives discussed here demonstrate that the climate crisis is foremost a social crisis. They stress that adaptation measures must therefore take questions of justice, concerning distribution, participation and recognition, into account. Instead of solely depicting how environmental injustices are aggravated by climate change, authors like Octavia Butler, envision ways that could lead to more viable futures and maintain that large-scale societal transformation is key.[30] Her book, *Parable of the Sower*, but also Zeitlin's *Beasts of the Southern Wild*, notably, also extend their justice claims beyond human relationships to non-human realms. Doing justice to nature here explicitly goes along with doing justice to humans. But while climate justice implies more sustainable ecological behavior as well as more just social practices, most climate change fictions prioritize the latter.[31] The role of the human, however, is highlighted not because nature is devalued, but precisely in order to show how humankind and nature are intricately related. The impacts of climate change are thus not just some meteorological events "out there," but profoundly affect the social setup of society. The climate change fictions discussed in this chapter render this relationship visible and thereby make the abstract and distant more easily apprehensible. Nixon has convincingly argued that narrative imagining is thereby "making it [the unapparent] accessible and tangible by humanizing drawn-out threats inaccessible to the immediate senses" (Nixon 15). By illustrating the impact of global warming on future generations as well as on people in different social, economic, and ethnic contexts, and by exposing in detail the injustice aggravated by climate change, fiction ultimately allows for readers and viewers to generate empathy, which may be crucial for taking action.

Notes

1. David Schlosberg argues that environmental justice is often based on Rawls's ideas of distributive justice (justice as fairness), and thus a focus on "a conception of justice defined solely as the distribution of goods in a society, and the best principles by which to distribute those goods" (3). The carbon debt approach is also based on distributional justice patterns.
2. In 2009, several developing countries submitted a proposal to the UNFCCC, which called for the payment of adaptation costs as well as emission debts by developed countries. The idea of "climate debt" has since also been employed by various NGOs and social movements as the only just mitigation measure to climate change (see also http://climate-debt.org/).
3. It should be noted, however, that the carbon trading structures that have been put in place so far have met significant resistance because they have failed to show their effectiveness of reducing CO_2 emissions. Furthermore, it may be argued that while climate debt ostensibly addresses issues of injustice, it actually functions like the selling of indulgences: affluent countries have the means to redeem themselves for the pollution and injustices that they are causing. In addition, the Clean Development Mechanism, which attempts to account for aspects of climate debts and the right of developing countries to economic development, has been losing in importance ever since the deflation of emission certificate prices and the failure to arrive at a binding post-Kyoto agreement (see Kreibich and Fechner).

4. Fiskio also draws on this quote to illustrate how the narrative of the lifeboat is based on the idea that human behavior solely follows self-interest and denies the possibility of the creation of communities.
5. In Chap. "Scaling Climate Change: The Transformation of Place in Climate Change Fiction", I analyzed how climate change fiction deals with the spatial dimension of climate change—that is, how it represents the phenomenon's globality. I argue that it is important to take this global dimension and its effects on place and belonging into account in order to understand climate change. By emphasizing the importance of the local dimension for climate justice, I do not mean to contradict my own earlier argument, but rather I am trying to problematize the difficulty of privileging either a sense of planet over a sense of place or the other way around. Sarah Ray's elaborations on Massey's "progressive sense of place" are helpful here because they go beyond place-space dualisms, refuse the local determinism as much as global openness, and point to the shifting glocal "power geometry" of people's relation to space (28). An analysis of these power networks—global *and* local—can make a valuable contribution to discussions about climate justice.
6. Krznaric, however, also points to the generic boundaries that may restrain blockbuster movies from providing an intimate perspective on the characters. As a successful example of empathic imagining, he names the docu-drama *The Age of Stupid*.
7. In *Poetic Justice*, Nussbaum elaborates on the importance of literature: "The literary imagination is part of a public rationality … In fact, I defend the literary imagination precisely because it seems to me an essential ingredient of an ethical stance that asks us to concern ourselves with the good of other people whose lives are distant from our own" (xvi). Drawing on various novels, Nussbaum is especially interested in the sense of life these texts embody. She also emphasizes literature's ability to generate emotions. Emotions are crucial in her political project because they get "people to think larger thoughts and recommit themselves to a larger common good" (*Political Emotions* 3).
8. In her article on environmental justice literature, Julie Sze explains that so far, sociological analyses dominate environmental justice discourse. She calls for ways to expand this "prism of statistics" and points to the importance of the cultural realm in making experiences of living with environmental racism more accessible (163–166). Sze, furthermore, emphasizes that climate justice is not just a "political

movement concerned with public policy ... [but also] a cultural movement interested in issues of ideology and representation" ("From Environmental Justice" 163). Though in the context of illness narratives, Priscilla Wald has also emphasized the power of literature and the cultural realm "to promote or mitigate the stigmatizing of individuals, groups, populations, locales (regional and global), behaviors, and lifestyles (3). In analyzing the key themes raised in these texts, literary scholars can make a valuable contribution to the discussion of environmental injustices.

9. The term "resilience" is frequently employed in climate change discourse and refers to the adaptability of a system to change. Resilient adaptation is often framed as an attempt to maintain a certain status quo with the help of technological or economic measures (see also Pelling 3). In the ecological use of the term, resilience also refers to the positive characteristics of a particular system of being resistant to disturbances. In literary studies, Murphy also points to the positive characteristics of resilience. He illustrates how the characters in several novels "cope, adapt, survive" and while no triumph is complete, "partial achievements form the basis for ongoing and productive practices that give rise to cultural continuity" ("Community Resilience" 150).
10. This resonates with environmental justice research, which has shown that communities of color are disproportionately affected by environmental pollution. Adamson et al. point to the important work by Reverend Benjamin Chavis on racial discrimination in environmental policy making, which demonstrated that poor and ethnic communities suffer disproportionate health risks (4). While Lolo's ethnic background is not further specified, the reasons he faces discrimination in Bacigalupi's short story are due to being his lower class and being "out of state," meaning outside of California.
11. By drawing on Rebecca Solnit's work, Janet Fiskio explains that depictions of human reactions to disaster are based on two conflicting assumptions about human nature: they either follow the idea of Social Darwinism and depict humans as essentially selfish—Fiskio terms this the narrative of the "lifeboat" or they believe in the altruistic nature of humankind, which Fiskio coins as the narrative of the "collective" (Fiskio, "Apocalypse" 14). Bacigalupi's short story takes the selfish impulse of humankind as a starting point, but at the same time critically exposes the consequences of such behavior.

12. Reports by the Western Water Assessment Center show that "The Colorado River provides water to 40 million people in seven western states and two countries and to 5.5 million irrigated acres" (Deems et al.). As climate change increasingly puts stress on the water flow of the river, issues of water rights gain in importance. Yet, scholars have pointed to a dearth of research in this regard while acknowledging that changes in the river's runoff will have a significant impact on water rights and "water right holders would likely have difficulty obtaining their decreed water rights" (Shapiro et al. 1).
13. The environmental justice scholar Robert Bullard identifies three categories in order to systematically examine the exposure of communities of color to environmental risks: procedural, geographic, and social equity.
14. After a shortage in vegetable harvest due to the internment of Japanese-American farmers, the US government encouraged citizens to contribute to the war effort by planting vegetables on their own properties. Without further acknowledging the injustices of the Japanese-American internment, Stein uses this war analogy to stress the dramatic effects that climate change will have on "home territory."
15. Even though Stein's novel explores the social dimensions of climate change, it should be noted that it does not account for race or gender differences.
16. I am aware that singular weather events such as Hurricane Katrina cannot be traced back to a long-term phenomenon like climate change. However, scientists point out that global warming has led to conditions in which the intensity and frequency of tropical storms is likely to increase (*IPCC Summary for Policy Makers 2013*, 5). In this chapter, therefore, I read Katrina as an example of extreme weather events that occur as a result of climate change.
17. As anthropologist Anthony Oliver Smith has argued, natural catastrophes are intricately related to the social processes in which they not only occur but which they also often disrupt. He explains, "In this sense a disaster is symptomatic of the condition of a society's total adaptational strategy within its social, economic, modified and built environments" (Oliver-Smith 25).
18. Already before the hurricane, New Orleans had a poverty rate of 23 percent. This poverty rate was exacerbated with the storm, since Katrina negatively affected those neighborhoods of the city that had a

particular high concentration of poverty and lacked adequate drainage systems, infrastructure, and housing, such as the Ninth Ward (Dyson 5–12).

19. Giroux draws here on the notion of biopolitics outlined by Foucault, who described it as a shift in the workings of power away from disciplining citizens toward regulating their very bodies. The ultimate goal is to foster life through continuous corrective bodily mechanisms (178). The power to produce and control life, as Giroux argues, however, is not indifferent to social structure, let alone to racial inequalities (181).
20. The media accounts in the aftermath of Katrina reported on lootings, murder and rape allegedly committed by the black population of New Orleans. Shortly after, the Governor's order was issued that allowed soldiers to kill looters. Soon military troops dominated the scene, underscoring the military character of the operation.
21. Clearly, it is questionable why the filmmakers decided to let Quvenzhané Wallis run around in underwear for a great part of the movie. Furthermore, hooks correctly criticizes the movie's stereotypical portrayal of the black father who is violent and alcoholic. However, *Beasts of the Southern Wild* does not romanticize the poor living conditions in Bathtub, but rather depicts the struggles of Hushpuppy's father who is desperately trying to build a future for his daughter in a world devastated by climate change.
22. The mythic tale about the aurochs depicted in the film resonates with the actual story of the aurochs, a kind of wild cattle that ranged in the early Holocene, and due to its domestication by humans, came to be extinct. The aurochs are thus not only a myth, but allude to humankind's early destructive path, which began with sedentariness and the rise of agriculture.
23. Schlosberg explains that with the 1990s, justice scholars turned to examine injustice as more than just maldistribution. He argues that recognition becomes a central concern as important justice theorists, such as Young and Fraser, "contend that a *lack* of recognition in the social and political realms, demonstrated by various forms of insults, degradation, and devaluation at both the individual and cultural level, inflicts damage to oppressed individuals and communities in the political and cultural realms" (14).
24. At various points Hushpuppy remarks on the interconnectedness of the universe. When the storm is approaching the Bathtub, Hushpuppy

hears the icebergs crashing into the sea. At an earlier point, she recognizes her animals and even humankind to be part of the "buffet of the universe" and tellingly observes, "The whole universe depends on everything fitting together. If one piece busts, even a small piece, the entire universe will get busted" (00:21:37–00:21:39).

25. While the novel is full of references to the Bible, it often points to the Bible's insufficiencies and people's narrow-minded interpretations of it. In Butler's second book, *Parable of the Talents,* the critique or Christianity is even more prevalent.
26. Though engaging with other text examples, Murphy argues that this community-building impulse constitutes an important aspect of resilience. Still, I would argue that Butler portrays more than just resilient communities, because with the spreading of the Earthseed religion, she ultimately calls for the transformation of the entire society and a radical change in the underlying belief system.
27. In analyzing Lauren's assumed difference from the bodily norm, Fiskio insightfully draws on disability studies and argues that in the novel hyperempathy syndrome serves "to revalue the possibilities of corporeal differences" ("Apocalypse" 29). According to the critic, Butler's story counters the idea of the disabled body as something abnormal and instead "new possibilities for alliance are sketched across the boundaries of corporeal differences" ("Apocalypse" 32).
28. Levinas's argument is especially interesting for environmental humanities because it counters the widely promoted idea that proximity to other beings (human and non-human) as well as a close attachment to the land is a prerequisite for moral responsibility.
29. For an extensive analysis of how *Parable of the Sower* falls into the tradition of the neo-slave narrative see Sylvia Mayer's article on genre and environmentalism.
30. By foregrounding issues of justice in a climatically changed world, the texts analyzed in this chapter offer a reflection on humankind and its capability to communally face this crisis or not.
31. Addressing ecological as well as environmental justice poses another challenge for climate change fiction authors. Heise's concept of eco-cosmopolitanism also touches on the problematic of extending ethical consciousness to the human as well as the non-human.

Conclusion: Climate Change Fiction and the Introduction of New Genres in Environmental Crisis Discourse

When Friedrich Christian Delius published his dissertation *Der Held und sein Wetter* in the 1970s, which turned into one of the best-selling works of German literary studies, the relationship between author, hero and weather still seemed stable and well-defined.[1] Analyzing several nineteenth century German novels (of the so-called *Bürgerlicher Realismus/bourgeois realism*), Delius intriguingly explains how writers used the weather as a literary device and thereby reveals the hidden ideologies of a genre based on the premises of realism. The critic furthermore explores the purpose of employing certain weather phenomena while trying to answer questions such as why does it always rain when the story comes to a tragic close, why is the arrival of the antagonist accompanied by the storm and what weather favors or challenges the hero, but, nevertheless, allows him to succeed. In addition to the textual analyses, Delius also shows how some weather passages challenge readers' expectations while others work to affirm certain societal constellations. As a reviewer at the time summarizes: "The fictitious weather is part of an emotionally charged nature and resonates with the lifestyle and values of the bourgeois heroes at the time" (Merker, my translation).[2] The weather thus functioned as a very effective and easily deployed literary tool for realist novels but also later works. Today, however, as Wolf Haas pointedly argues in his preface to Delius's reprint, the weather no longer easily serves as accomplice to the hero of the story. Neither background setting nor "atmospheric rendition

A. Mehnert, *Climate Change Fictions*,
DOI 10.1007/978-3-319-40337-3_8

of services for heroes of any kind" (Delius 5, my translation), the weather itself has, instead, turned into a primary character of the narration.[3]

At the beginning of the twenty-first century, anthropogenic climate change is undisputable. As humans are altering the Earth's climate, most notably the relationship between nature and culture has become unsettled. Ideas of nature as something untouched, pure and wild are no longer suitable to describe the complex intertwining of human and nature in the Anthropocene. Nature is neither a space apart from the human nor completely dominated by humankind—even though humans' intervention into the planetary systems is severe. In this context, also the weather is no longer mere backdrop to human interaction but prime site of attention as well as contestation. As the climatically changed planet enters the story not as an "extra" but as another prominent character, humans in general and writers in particular have to renegotiate their own position with regard to "the weather."

While the works discussed in this book are very diverse, it is precisely this negotiation process that lies at their core, and, ultimately, constitutes their common characteristic. Climate change fiction can then be understood as texts explicitly dealing with anthropogenic climate change. In contrast to the works that Delius discusses, climate change fiction does not employ weather-related phenomena either as setting or for comfort to the characters, but climate change forms an inherent part of the narration. It is important for the characters' interaction among each other as well as their interaction with the planet. Yet, in order to allow for the climatically altered planet to fully enter the story, writers of climate change fiction have to overcome several representational and conceptual challenges that global warming poses. While scientists overwhelmingly agree that global temperatures are rising due to increasing greenhouse gas emissions, the exact ramifications of this transformation are difficult to define. Moreover, the global scale, the disconnect between cause and effect, its slow agent, complexity and uncertainty turn climate change into a very elusive issue that demands innovative approaches for its representation. Yet, as I have attempted to show, writers engaging with climate change are increasingly taking on these challenges, thereby shaping our very understanding of this unprecedented phenomenon.

Steven Amsterdam's *Things We Didn't See Coming* and Barbara Kingsolver's *Flight Behavior* provide innovative narrative means to engage with the spatial dimension of climate change. They illustrate the deterritorializing effects of global warming as attachments to place are challenged and forms of belonging and inhabitation questioned. Offering an "eco-cosmopolitan" (Heise) vision, Kingsolver ultimately emphasizes that natural and cultural processes shape one another in disregard of local or national borders– especially in times of climate change. An understanding of this planetary connectedness is then crucial to address the climate crisis. Amsterdam, on the other hand, postulates that the deterritorializing effects of global warming are so far-reaching that we need to think about climate change in terms of "global uprootedness." Accordingly, any place attachment is futile, and humans are turned into solitary passengers without any fixed habitats. Both works finally show that the planetary transformations that climate change entails, require thinking beyond place-based routines and opening up to global perspectives.

While Amsterdam and Kingsolver engage in questions of place, Boyle and McNeil turn to the temporal dimension of climate change. Their books, *A Friend of the Earth* and *The Ice Lovers*, allow for a discussion of how environmental timescapes are rendered visible in fiction. Especially McNeil's work, with its focus on the Antarctic, reveals those environmental temporalities that often remain imperceptible. It thereby not only articulates the "slow violence" (Nixon) of climate change but also underscores the inseparability of natural dynamics and human timescales. Moreover, pointing to the dangers of short-term thinking, the climate change fictions discussed in Chap. "Reimagining Time in Climate Change Fiction" create the narrative relationships between past, present and future while exposing the potential dangers that a bifurcation of time otherwise harbors. Boyle employs an interesting narrative structure to counter the loss of perspective and to reconstruct a timescape perspective that engages in the temporal aspects of living in times of climate *change*.

The complexity, invisibility and future dimension of climate change raise several questions of uncertainty, which have either fed into general societal ignorance or confusion about the issue, and, as explained in Chap. "Climate Change Fictions in Context: Socio-Politics, Environmental Discourse and Literature", served to support conservative political projects. In trying to analyze how the risks and uncertainties of climate change are communicated and critically dealt with, Chap. "ClimateCultures in Kim Stanley Robinson's *Science in the Capital* Trilogy" draws on Nathaniel Rich's *Odds Against Tomorrow* and juxtaposes parts of it with Michael Crichton's *State of Fear*. Highlighting what kinds of risk need to be contemplated and taken seriously while critically exposing the "commodification of contingency" (Lobo-Guerrero), Rich offers a path through this complex web of climate uncertainties and criticizes socio-political passivity in the face of it.

As discussed in the subsequent chapter, Kim Stanley Robinson's climate change fiction trilogy broaches the issue of how climate change defamiliarizes our understanding of nature by challenging Western key dichotomies such as the nature-culture divide. The trilogy foregrounds that the human is always already part of a hybrid and co-produced nature. Consequently, rather than trying to preserve a separate pristine nature, the important question is how to responsibly terraform a planet constantly in flux. Robinson finally envisions one possible story of adaption, one, which acknowledges that human and nature are no longer separate entities in the Anthropocene.

While Robinson's trilogy critically asks what "natures" humans are willing to give up, the climate change fictions discussed in the last part of this book remind readers exactly of the injustices that this giving up may involve. Paolo Bacigalupi's "The Tamarisk Hunter," Benh Zeitlin's *Beasts of the Southern Wild*, Octavia Butler's *Parable of the Sower* and Dana Stein's *Fire in the Wind* depict those issues so far underrepresented in scientific graphs and illustrations but crucial to the climate change debate. These cultural texts employ climate change to tease out questions of justice and responsibility. Climate change fiction thereby gives insight into the ethical and socio-political ramifications of this unparalleled environmental crisis.

In grappling with these challenges, the books chosen for analysis all, despite their diversity, inevitably show a common characteristic: anthropogenic climate change emerges as a central element for the narrative and is made visible in its multiple facets to the reader. This points to a fairly open thematic definition of what could be read as a newly emerging genre called climate change fiction. Defining climate change fiction, however, in terms of a characteristic form or technique is more complicated. As the analyses in this book have shown, the texts range from science fiction and magical realism to novels and short stories and draw on various literary modes such as the ironic, the didactic, and the dystopian. Engaging in the manifold representational challenges of climate change pushes writers toward literary innovation and to move beyond genre conventions. While some works, like Amsterdam's *Things We Didn't See Coming,* introduce new styles such as the stop-start pattern, others, like Robinson's trilogy, blend science fiction with utopian realism. Again others fuse magical elements with realism, as in *Beasts of the Southern Wild*. It seems then that hybridized literary forms grasp best the complexity of climate change.

Though not subject of discussion in this book, it is important to note that non-fiction dealing with climate change also makes use of different literary styles and no longer strictly follows genre definitions of non- fiction; some works could even be read as a blend of non-fiction with science fictional elements. Other works, like James Hansen's *Storms of our Grandchildren*, prominently engage in apocalyptic rhetoric and draw on a dystopian tone. The fusion of stylistic elements thus plays an important role across genre categories.

Even though there may not be one specific genre most suitable for climate change fiction, a discussion of the differences in genres employed is, nevertheless, important. Agreeing with Garrard, "An author's choice of genre in writing about climate change is crucial: it makes some sorts of action possible and others impossible" ("The Unbearable Lightness" 9). Garrard, for example, refers in this context to the ambivalent use of comic allegory in McEwan's *Solar*, which according to him is too accepting of the denial it seeks to expose and thus limits the story's moral value. So the genre and narrative framework of a cultural text guides readers' understanding and response to the issues with which the text is concerned.

Indeed, and as explained in Chap. "Climate Change Fictions in Context: Socio-Politics, Environmental Discourse and Literature", environmental discourse has so far often relied on apocalypticism in order to persuade people of the urgency to take action. Rachel Carson or Paul Ehrlich's well-known non-fiction writings, but also several works of fiction published at the time, prove a case in point. Acknowledging the importance of these works in putting the environment at the center of attention, Frederick Buell points out, "Together, they distinctively and creatively gave force to the feeling that the environmental problems of the day amounted to apocalypse—to rupture, world-end and a last judgment on humanity" ("A Short History" 18). Moreover, they revealed that this environmental apocalypse was human-made. While also today eschatological rhetoric prominently figures in media as well as non-fiction accounts about the future ramifications of climate change, the climate change fictions analyzed here for the most part reinvent apocalypse or refrain from drawing on it at all.

Engaging in the numerous challenges that climate change presents, the employment of apocalypse, of a revelatory moment of the world's end, does not seem adequate to grasp the slow and "unspectacular" character of climate change, which does not manifest in a singular moment but on multiple scales and scopes. So instead of playing out in one catastrophic event, global warming produces numerous and continuous crises. Similar to the diffuse character of its effects, there is not one single cause or culprit for climate change. Whereas apocalypse relies on moral dichotomies and thus on clearly defined enemies and antagonists, in climate change fiction these are not easily identified.[4] Though it is clear that the climatic changes depicted in the novels are the consequences of human behavior (anthropogenic), the works do not present a last judgment scenario punishing all humans. Even in books such as *Odds Against Tomorrow*, in which a huge flood destroys a great part of New York, Nathaniel Rich criticizes apocalyptic rhetoric for its fear-mongering effect. Though his protagonist acknowledges, "Problems would arise that he could not anticipate … The weather would only become increasingly erratic" (Rich 286), Rich challenges the paralysis induced by fear and in particular scare scenarios throughout his novel. Instead of portraying a singular revelatory catastrophe, climate change fiction, such as Rich's novel, focuses on how humans adapt and survive in a world continuously faced with numerous climate change crises. By suggesting that different futures—though never one that is risk-free—are nevertheless possible, the texts underscore the current moment of *krisis* (Ancient Greek for "decision and choice"). Not necessarily apocalyptic but risk narratives, climate change fictions still convey the sense of urgency while exploring ways of adaptation and mitigation. Climate change fiction hereby already hints at the diversification of narrative frameworks used to convey contemporary environmental crises. While my chapter analyses elaborate in more detail how the texts reinvent apocalypse, this does not constitute the sole focus of this book project. It has, however, been fruitful to consider climate change fiction in the context of "risk narratives," especially in relation to Buell's ide of contemporary "domestication of crisis." Since climate change fiction remains a highly diverse genre, future critical work engaging in a detailed analysis of the narrative techniques, genres and modes employed in climate change fiction would make a valuable contribution to the discussions of this newly emerging field.

In envisioning the future, climate change fiction also allows readers to reflect on and critically assess the social ideas it projects.[5] Octavia Butler, for example, most prominently argues for social transformation. She emphasizes the importance of respect for difference and distributional equity in order to create more just societies. While in *Parable of the Sower* communities based on social justice are successfully created despite of severe climatic changes, Nathaniel Rich and Steven Amsterdam are more pessimistic. Skeptical of the altruistic nature of humankind, they portray that social disintegration is inevitable in a climatically changed world, and thus self-reliant survivalism is the only adequate response. Both writers, however, expose the negative and undesirable aspects of such behavior.[6] Again others like Kim Stanley Robinson call for a communal effort in society to change the current political and economic system and to thus enable sustainable life on earth.

In this regard, the national socio-political context laid out in Chap. "Climate Change Fictions in Context: Socio-Politics, Environmental Discourse and Literature" plays an important role finding resonance in the texts. Rich denounces a growing industry in the United States that encourages profit making from disaster; Robinson is critical of American right-wing politics for their liberal market policies and denial of climate change, and Barbara Kingsolver criticizes the role of the Religious Right in endorsing climate change skepticism. As Axel Goodbody has convincingly argued, in *Flight Behavior*, "Religion emerges as a form of denial" ("Risk, Denial" 53). The people of Feathertown ignore the strange new migration route of the butterflies since to them weather is "the Lord's business" (Kingsolver 261). Moreover, Kingsolver addresses current political inaction toward issues of economic disparity as well as inequality of educational opportunity in her novel and shows that these are important aspects that need to be taken into consideration in the discussions about climate change.

Not only resonating the socio-political context during the time of their creation process, many climate change fictions also root their narratives in a very recognizable American setting. They thus explicitly counter the idea that the United States is an unaffected bystander to climate change. From "Los Andiegoles" to New York, from Louisiana to Iowa, the texts pin the effects of global warming on the American map. A comparative analysis with climate change fiction from different national contexts as part of future research on climate change fiction would surely provide interesting insights in order to understand whether there is a national—cultural element in risk-perception and to explore potential international differences in the representation of climate change.[7]

Reading this turn toward the national in American climate change fiction as a response to the fact that climate change remains a highly contested issue in the United States, some of the works thereby reveal their didactic impulses (authors like Robinson or Stein have also openly stated this as their intention). In his analysis of *EisTau* by Ilija Trojanow, a work of German climate change fiction, Goodbody has insightfully argued that Trojanow's literary attempt to prompt readers to take action against climate change, comes dangerously close to preaching (100). Thus, conveying openly the didactic intention of their works, writers run the risk of further estranging readers. Yet, interestingly, as international political action to mitigate climate change is only slowly advancing and the issue of climate crisis is becoming more pressing, numerous newspaper articles were published conjuring the persuasive power of literature as well as its interventionist, educational and political nature (see Korman, Macfarlane, Kramb). Literature, seemingly, is seen as the last resort to communicate the urgency of environmental crisis and prompt action. However, Greg Garrard cautions especially ecocritics to read literature as a means of persuasion: "Ecocriticism is not the literary critical department of the IPCC" ("The Unbearable Lightness" 17), he writes. Agreeing with Garrard, it is problematic and difficult for ecocritics to convincingly argue that literature has the persuasive power to prompt readers to take action against climate change. Still, literature's ability to enlighten and to compellingly convey ideas, especially in the context of global warming, is not to be neglected.

The analyses in this book have demonstrated literature's particularly well-suited means to envision this elusive and abstract phenomenon called climate change. *Climate Change Fictions* also aims to make a contribution to the debates about how environmental crisis and global warming in particular can be communicated. As writers are overcoming the challenges of representing climate change, they do not only offer narratives that provide insight into the personal and emotional contours of climate change, but also shape our very idea and understanding of this invisible crisis. Climate change is no longer "out there," but turned into an intricate part of the characters' as well as readers' reality. Resonating this shift, climate change has also, though still as a secondary issue, found its way into several contemporary American books. Since a world without climate change has become difficult to imagine and global warming increasingly plays a role in the fictional realities of mainstream literature, not only ecocritics but also other literary scholars will be prompted to open up their discussions beyond disciplinary boundaries and to allow for new critical climates to emerge.

Notes

1. Unfortunately, the book has not been translated into English. The title of the book can be translated as "the hero and his weather."
2. Merker writes in the German original: "Das fiktive Wetter als Teil der emotional beladenen Natur wird als Stimmungsmacher für bürgerliche Helden, Lebensweisen und Werturteile gezeigt."

3. Haas writes in the German original: "Vorbei ist es mit der dienenden Rolle des Wetters als Hintergrundmythos, mit der atmosphärischen Dienstleistung für Helden aller Art" (Delius 5).
4. It is important to point out that the argument I present here applies to the novels discussed in this book. Genre conventions continue to exercise significant power over a writer's engagement in a specific topic. Climate change thrillers such as Matthew Glass's *Ultimatum*, for example, heavily rely on spectacular showdowns and easily identified enemies, and thus show that the depth of the engagement in climate change also depends on the openness of the writer to literary innovation.
5. As the analyses have shown, climate change fiction also reveals several underlying social imaginaries. A detailed discussion of these social imaginaries, and of the specific political and economic projects they endorse, would certainly also provide answers to the question "what we really want to achieve for humanity" ("Four Meanings" 55), which Mike Hulme has notably raised and addressed. Future research on climate change fiction will hopefully take this question into account.
6. See also my discussion of Janet Fiskio's distinction between "lifeboat" and "collective" narratives in Chap. "Representing the Underrepresented: Climate Justice and Future Responsibilities in Climate Change Fiction".
7. German Climate Change fiction, for example, often employs a more global setting and does not necessarily focus on Germany as the context in which the story unfolds. See also my and Gregers Andersen's analysis in "Der Held und das Wetter: Literatur und Klimawandel."

Bibliography

Adam, Barbara. "Re-Vision: The Centrality of Time for an Ecological Social Science Perspective." *Risk, Environment, and Modernity: Towards a New Ecology*. Ed. Scott Lash. London: Sage Publications, 1996. 84–103. Print.

———. *Timescapes of Modernity: The Environment and Invisible Hazards*. London: Routledge, 1998. Print.

Adamson, Joni, Mei Mei Evans, and Rachel Stein (eds). *The Environmental Justice Reader: Politics, Poetics, & Pedagogy*. University of Arizona Press, 2002. Print.

Agamben, Giorgio. "The Time That Is Left." *Epoche* 7.1 (2002): 1–14. Print.

Alaimo, Stacy. *Bodily Natures: Science, Environment, and the Material Self*. Indiana University Press, 2010. Print.

Amsterdam, Steven. *Things We Didn't See Coming*. London: Vintage Books, 2011. Print.

Anderson, Benedict R. *Imagined Communities: Reflections on the Origin and Spread of Nationalism*. London; New York: Verso, 1991. Print.

An Inconvenient Truth. Dir. Davis Guggenheim. Stars. Al Gore. Lawrence Bender Productions, 2006. Film.

Appadurai, Arjun. "Disjuncture and Difference in the Global Cultural Economy." *Theory, Culture & Society* 7.2 (1990): 295–310.

———. *Modernity At Large: Cultural Dimensions of Globalization*. U of Minnesota P, 1996. Print.

Archer, David, and Stefan Rahmstorf. *The Climate Crisis: An Introductory Guide to Climate Change*. Cambridge: Cambridge UP, 2010. Print.

Armbruster, Karla, and Kathleen Wallace. *Beyond Nature Writing: Expanding the Boundaries of Ecocriticism*. UP of Virginia, 2001. Print.

A. Mehnert, *Climate Change Fictions*,
DOI 10.1007/978-3-319-40337-3

Augé, Marc. *Non-Places: Introduction to an Anthropology of Supermodernity.* Trans. John Howe. London; New York: Verso, 1995. Print.

Bacigalupi, Paolo. "The Tamarisk Hunter." *I'm With the Bears.* New York: Verso, 2011. 171–191. Print.

———. *The Windup Girl.* Nightshade Book, 2010. Print.

Bateson, Gregory. *Steps to an Ecology of Mind: Collected Essays in Anthropology, Psychiatry, Evolution and Epistemology.* U of Chicago P, 2000. Print.

Bawarshi, Anis. "The Ecology of Genre." *Ecocomposition: Theoretical and Pedagogical Approaches.* Ed. Christian Weissner and Dobrin Sidney. Albany: State University of New York Press, 2001. Print.

Beasts of the Southern Wild. Dir. Zeitlin, Benh. Perf. Quvenzhané Wallis, Dwight Henry. Fox Searchlight, 2012. Film.

Beck, Ulrich. *Ecological Politics in an Age of Risk.* Cambridge; Malden, MA: Polity Press, 2002. Print.

———. *Risk Society: Towards a New Modernity.* London: SAGE, 1992. Print.

———. *World at Risk.* Malden, MA: Polity, 2009. Print.

———. *World Risk Society.* Malden, MA: Blackwell Publishers, 1999. Print.

———. "World Risk Society and Manufactured Uncertainties." *Iris* 1.2 (2009): 291–299. Print.

Beck, Ulrich, Anthony Giddens, and Scott Lash. *Reflexive Modernization: Politics, Tradition and Aesthetics in the Modern Social Order.* Cambridge, UK: Polity Press, 1994. Print.

Bennett, Jill. *Living in the Anthropocene.* Kassel: Documenta 13, 2011. Print.

Bennett, Michael. "From Wide Open Spaces to Metropolitan Places. The Urban Challenge to Ecocriticism." *Interdisciplinary Studies in Literature and Environment* 8.1 (2001): 31–52. Print.

Bianco, Nicholas, Franz Litz, and Thomas Damassa. "Reducing Greenhouse Gas Emissions in the United States Using Existing Federal Authorities and State Action." *World Resources Institute.* July 2010. Web. 10 Feb. 2014.

Bould, Mark, and Sherryl Vint. *The Routledge Concise History of Science Fiction.* New York: Routledge, 2011. Print.

Boyer, Christine. *The City of Collective Memory.* Cambridge, Mass.: MIT Press, 1996. Print.

Boykoff, Jules, and Maxwell Boykoff. "Journalistic Balance as Global Warming Bias: Creating Controversy Where Science Finds Consensus." *Global Environmental Change Part A* 14.2 (2004): 125–136. Print.

Boykoff, Maxwell T. *Who Speaks for the Climate?: Making Sense of Media Reporting on Climate Change.* Cambridge University Press, 2011. Print.

Boyle, Tom Coraghessan. *A Friend of the Earth.* London: Bloomsbury Publishing, 2004. Print.

———. *When the Killing's Done.* London: Bloomsbury, 2012. Print.

Brandt, Stefan, Winfried Fluck, and Frank Mehring. "Introduction. Transcultural Spaces." *REAL 26. Transcultural Spaces: Challenges of Urbanity, Ecology and the Environment*. Ed. Stefan Brandt, Winfried Fluck, and Frank Mehring. Tübingen: Narr Verlag, 2010. Print.

British Antarctic Survey. "Antarctica and Climate Change". UK: National Environmental Research Council, 2009. Web. 6 Feb. 2013.

Brooks, Rosa. "Our Homegrown Third World." *LA Times. Los Angeles Times* 7 Sept. 2005. Web. 21 Dec. 2013.

Buell, Frederick. "A Short History of Environmental Apocalypse." *Future Ethics: Climate Change and Apocalyptic Imagination*. Ed. Stefan Skrimshire. London: Continuum Inter. Publis., 2010. 13–36. Print.

———. *From Apocalypse to Way of Life : Environmental Crisis in the American Century*. New York: Routledge, 2004. Print.

Buell, Lawrence. "Nature and City: Antithesis or Symbiosis ?" *REAL 26. Transcultural Spaces: Challenges of Urbanity, Ecology and the Environment*. Ed. Stefan Brandt, Winfried Fluck, and Frank Mehring. Tübingen: Narr Verlag, 2010. 3–20. Print.

———. *The Environmental Imagination: Thoreau, Nature Writing, and the Formation of American Culture*. Harvard UP, 1996. Print.

———. *The Future of Environmental Criticism : Environmental Crisis and Literary Imagination*. Malden, MA: Blackwell Pub., 2005. Print.

———. "Toxic Discourse." *Critical Inquiry* 24.3 (1998): 639–665. Print

———. *Writing for an Endangered World: Literature, Culture, and Environment in the U.S. and Beyond*. Harvard UP, 2003. Print.

Bulkeley, Harriet. *Cities and Climate Change*. New York: Routledge, 2013. Print.

Bullard, Robert. "Poverty, Pollution and Environmental Racism: Strategies for Building Healthy and Sustainable Communities." *Environmental Justice Resource Center*. July 2002. Web. 20 Jan. 2014.

Butler, Octavia. *Parable of the Sower*. New York; Boston: Grand Central Publishing, 1993. Print.

———. *Parable of the Talents*. New York: Open Road Media, 2012. Kindle.

Canclini, Nestor Garcia. *Hybrid Cultures: Strategies for Entering and Leaving Modernity*. Trans. Christopher L. Chiappari and Silvia L. Lopez. U Of Minnesota P, 2005. Print.

Carson, Rachel. *Silent Spring*. 1962. Boston: Houghton Mifflin, 2002. Print.

Chakrabarty, Dipesh. "Postcolonial Studies and the Challenge of Climate Change." *New Literary History* 43.1 (2012): 1–18.

———. "The Climate of History: Four Theses." *Critical Inquiry* 35 (Winter 2009): 197–222. Print.

Cho, Daniel K. "'When a Chance Came For Everything to Change': Messianism and Wilderness in Kim Stanley Robinson's Abrupt Climate Change Trilogy." *Criticism* 53.1 (2011): 23–55. Print.

Clark, Nigel. "Panic Ecology: Nature in the Age of Superconductivity." *Theory, Culture & Society* 14.1 (1997): 77–96.

Clark, Timothy. "Some Climate Change Ironies : Deconstruction, Environmental Politics and the Closure of Ecocriticism." *Oxford Literary Review* 1. 2010. 131–149. Print.

———. *The Cambridge Introduction to Literature and the Environment.* Cambridge University Press, 2011. Print.

Clifford, James. *Routes: Travel and Translation in the Late 20th Century.* Harvard University Press, 1997. Print.

Commoner, Barry. *The Closing Circle: Nature, Man, Technology.* New York: Alfred Knopf, 1971. Print.

Connor, Steven. *The Cambridge Companion to Postmodernism.* Cambridge University Press, 2004. Print.

Coole, Diana, and Samantha Frost, eds. *New Materialisms: Ontology, Agency, and Politics.* Duke University Press Books, 2010. Print.

Coste, Didier, and John Pier. "Narrative Levels." *The Living Handbook of Narratology.* 5 Sept. 2013. Web. 22 Jan. 2014.

Coupe, Laurence. *The Green Studies Reader: From Romanticism to Ecocriticism.* New York: Routledge, 2000. Print.

Crichton, Michael. *State of Fear.* Harper Collins, 2004. Print.

———. "The Role of Science in Environmental Policy-Making". U.S. Senate Committee on Environment & Public Works. 28 Sept. 2005. Web. 12 July 2013.

Cronon, William. "The Trouble with Wilderness; Or, Getting Back to the Wrong Nature." *Uncommon Ground: Rethinking the Human Place in Nature.* London: Norton, 1996. Print.

Csicsery-Ronay, Istvan. *The Seven Beauties of Science Fiction.* Middletown: Wesleyan UP, 2008. Print.

Cutter, Susan L. *Hazards, Vulnerability and Environmental Justice.* London: Earthscan, 2006. Print.

Davis, Angela Y. *Abolition Democracy.* New York: Seven Stories, 2005. Print.

Deems, J. S. et al. "Combined Impacts of Current and Future Dust Deposition and Regional Warming on Colorado River Basin Snow Dynamics and Hydrology." *Hydrology and Earth System Sciences* 17.11 (2013): 4401–4413. Print.

Deleuze, Gilles, and Felix Guattari. *Anti-Oedipus: Capitalism and Schizophrenia.* 1972. Trans. Robert Hurley, Mark Seem, and Helen Lane. 6th printing. Penguin Classics, 2009. Print.

Delius, F.C. *Der Held und Sein Wetter.* Göttingen: Wallstein Verlag, 2011. Print.

Dery, Mark. "Culture Jamming: Hacking, Slashing, and Sniping in the Empire of Signs." *Shovelware* 8 Oct. 2010. Web. 7 March 2012.

Dessler, Andrew E., and Edward A. Parson. *The Science and Politics of Global Climate Change: A Guide to the Debate.* Cambridge UP, 2006. Print.

Diamond, Jared. *Collapse.* Rev. Ver. New York: Penguin, 2011. Print.

Dillon, Michael. "The Biopolitics of Security." *The Routledge Handbook of New Security Studies.* London; New York: Routledge, 2010. 61–72. Print.

Dillon, Michael, and Luis Lobo-Guerrero. "Biopolitics of Security in the 21st Century: An Introduction." *Review of International Studies* 34.02 (2008): 265–292. Print.

Douglas, Mary, and Aaron Wildavsky. *Risk and Culture: An Essay on the Selection of Technical and Environmental Dangers.* Berkeley: U of California P, 1982. Print.

Dyson, Michael Eric. *Come Hell or High Water: Hurricane Katrina and the Color of Disaster.* Basic Books, 2006. Print.

Edelglass, William. "Rethinking Responsibility in an Age of Anthropogenic Climate Catastrophe." *Facing Nature: Lévinas and Environmental Thought.* Ed. William Edelglass, James Hatley, and Christian Diehm. Pittsburgh, Pa.: Duquesne University Press, 2012. 209–228. Print.

Edwards, Paul N. "Representing the Global Atmosphere." *Changing the Atmosphere: Expert Knowledge and Environmental Governance.* Ed. Clark Miller. Cambridge, Mass: MIT Press, 2001. 31–67. Print.

Egan, Jennifer. *A Visit from the Goon Squad.* Norwell, MA: Anchor, 2011. Print.

Ehrlich, Paul. *The Population Bomb.* Rev. Macmillan, 1971. Print.

Evernden, Neil. *The Natural Alien: Humankind and Environment.* 2nd ed. Toronto: U of Toronto P, Scholarly Publishing Division, 1993. Print.

Ewald, Francois. "The Return of Descartes's Malicious Demon: An Outline of a Philosophy of Precaution." Trans. Stephen Utz. *Embracing Risk, the Changing Culture of Insurance and Responsibility.* Ed. T. Baker and J. Simon. Chicago: U of Chicago P, 2002. 273–301. Print.

Faber, Daniel J. *The Struggle for Ecological Democracy: Environmental Justice Movements in the United States.* New York: Guilford Press, 1998. Print.

Feldman, M.B., and H.L. Hsu. "Introduction: Race, Environment, and Representation." *Discourse* 29.2-3 (2009): 199–214. Print.

Fisher, Susannah. "The Emerging Geographies of Climate Justice." *The Centre for Climate Change Economics and Policy. Working Paper No 94* (June 2012). Print.

Fiskio, Janet. "Apocalypse and Ecotopia: Narratives in Global Climate Change Discourse." *Race, Gender & Gender* 19.1-2 (2012): 12–37. Print.

———. "Rethinking Agrarianism, Place, and Citizenship." *American Literature* 84.2 (2012): 301–325. Print.

Flannery, Tim. *The Weather Makers : How Man Is Changing the Climate and What It Means for Life on Earth.* First American Edition. Atlantic Monthly Press, 2006. Print.

Fleck, Dirk. *Das Südsee Virus.* München: Piper, 2012. *Print.*

Flippen, J. Brooks. "Richard Nixon and the Triumph of Environmentalism." *American Environmental History.* Ed. Louis S. Warren. Malden, MA: Blackwell, 2002. 272–289. Print.

Flottum, Kjersti. "A Linguistic and Discursive View on Climate Change Discourse." *A. sp. Anglais de spécialité* 58 (2010): 19–37. Print.
Flow. For the Love of Water. Dir. Irena Salina. Group Entertainment, 2008. Film.
Food, Inc. Dir. Robert Kenner. Stars Michael Pollan. Magnolia, 2008. Film.
Forsyth, Tim. "Politicizing Environmental Science Does Not Mean Denying Climate Science Nor Endorsing It Without Question." *Global Environmental Politics* 12.2 (2012): 18–23. Print.
Foucault, Michel. *Security, Territory, Population. Lectures at the Collège de France, 1977-78.* New York: Picador, 2007. Print.
———. *Society Must Be Defended. Lectures at the Collège de France, 1975-1976.* New York: Picador, 2003. Print.
Franzen, Jonathan. *Freedom.* Macmillan, 2011. Print.
Fraser, Nancy. "Rethinking Recognition." *New Left Review* May/June (2000): 107–20. Print.
———. "Recognition without Ethics?" *Theory, Culture, and Society* 18. 2–3. (2001): 21–42. Print.
Furedi, Frank. "Precautionary Cultrue and the Rise of Risk Assessment." *Erasmus Law Review* 2.2 (2009): 197–220. Print.
Gaarder, Jostein. *2084-Noras Welt.* Trans. Gabriele Haefs. München: Hanser, 2013. Print.
Gabrys, Jennifer, and Kathryn Yusoff. "Arts, Sciences and Climate Change: Practices and Politics at the Threshold." *Science as Culture* 21.1 (2012): 37–41. Print.
Garrard, Greg. *Ecocriticism.* New York: Routledge, 2004. Print.
———. *Teaching Ecocriticism and Green Cultural Studies.* Palgrave, 2012. Print.
———. "The Unbearable Lightness of Green: Air Travel, Climate Change, and Literature." University of Oregon: N. p., 2013. 1–18.
Gersdorf, Catrin, and Sylvia Mayer. "Nature in Literary and Cultural Studies: Defining the Subject of Ecocriticism-an Introduction." *Nature in Literary and Cultural Studies: Transatlantic Conversations on Ecocriticism.* Ed. Catrin Gersdorf and Sylvia Mayer. Amsterdam: Editions Rodopi BV, 2006. 9–25. Print.
Giddens, Anthony. *Consequences of Modernity.* Oxford: Blackwell Publishers, 1991. Print.
———. *The Politics of Climate Change.* London: Polity, 2009. Print.
Giroux, Henry. "Reading Hurricane Katrina: Race, Class, and the Biopolitics of Disposability." *College Literature* 33.3 (2006): 171–196. Print.
Glasberg, Elena. "'Living Ice': Rediscovery of the Poles in an Era of Climate Crisis." *WSQ: Women's Studies Quarterly* 39.3-4 (2011): 221–246. Print.
Glass, Matthew. *Ultimatum.* Atlantic Books, 2010. Print.
Glotfelty, Cheryll, and Harold Fromm. *The Ecocriticism Reader: Landmarks in Literary Ecology.* U of Georgia P, 1996. Print.

Godfrey, Phoebe C. "Introduction: Race, Gender & Class and Climate Change." *Race, Gender & Class* 19.1-2 (2012): 3–12. Print.

Goodbody, Axel H. "Melting Ice and the Paradoxes of Zeno: Didactic Impulses and Aesthetic Distanciation in German Climate Change Fiction." *Ecozon@* 4.1 (2013): 92–102. Print.

———. "Risk, Denial and Narrative Form in Climate Change Fiction: Barbara Kingsolver's Flight Behavior and Ilija Trojanow's Melting Ice." *The Anticipation of Catastrophe.* Ed. Sylvia Mayer and Alexa Weik von Mossner. Heidelberg: Winter, 2014. 39–59. Print.

Gramelsberger, Gabriele. "Intertextualität und Projektionspotenzial von Klimamodellen." *Prophetie und Prognostik.* Ed. Daniel Weidner and Stefan Willer. München: Wilhelm Fink, 2013. 209–225. Print.

Gregg, Melissa, and Gregory J. Seigworth. *The Affect Theory Reader.* Durham: Duke University Press, 2010. Print.

Gregory, Derek, and John Urry. "Introduction." *Social Relations and Spatial Structures.* Macmillan, 1985. 1–8. Print.

Hansen, James. *Storms of My Grandchildren: The Truth about the Coming Climate Catastrophe and Our Last Chance to Save Humanity.* Bloomsbury Publishing, 2011. Print.

Haraway, Donna Jeanne. *Simians, Cyborgs and Women. The Reinvention of Nature.* New York: Routledge, 1991.

———. *When Species Meet.* Minneapolis: U of Minnesota P, 2007. Print.

Harvey, David. *The Condition of Postmodernity.* Malden, Mass.: Blackwell, 1990.

Heise, Ursula K. *Chronoschisms: Time, Narrative, and Postmodernism.* Cambridge: Cambridge UP, 1997. Print.

———. "Cultures of Risk and the Aesthetic of Uncertainty." *Scientific Cultures-Technological Challenges.* Ed. Klaus Benesch and Maike Zwingenberger. Heidelberg: Winter, 2009. 17–44. Print.

———. "Comparative Ecocriticism in the Anthropocene." *Komparatistik* 25 (forthcoming): 15–26. Print.

———. "Introduction: The Invention of Eco-Futures." *Ecozon@* 3.2 (2012): 1–10. Print.

———. "Lost Dogs, Last Birds, and Listed Species: Cultures of Extinction." *Configurations* 18.1-2 (2010): 49–72. Print.

———. "Teaching Ecocritical Theory." *Teaching North American Environmental Literature.* Ed. Laird Christensen. New York: Modern Language Association of America, 2008. 44–57. Print.

———. "The Hitchhiker's Guide to Ecocriticism." *PMLA* 121.2 (2006): 503–516.

———. "Toxins, Drugs, and Global Systems: Risk and Narrative in the Contemporary Novel." *American Literature* 74.4 (2002): 747–778. Print.

———. *Sense of Place and Sense of Planet: The Environmental Imagination of the Global.* New York: Oxford UP, 2008. Print.

Herzog, Arthur. *Heat*. Simon & Schuster, 1976. Print.

Hobday, Alistair. "Sliding Baselines and Shuffling Species: Implications of Climate Change for Marine Conservation." *Marine Ecology* 32 (2011): 392–403. Print.

hooks, bell. "No Love in the Wild." *NewBlackMan* 6 Sept. 2012. Web. 2 Jan. 2014.

Horn, Eva. "Der Anfang vom Ende. Worst-Case-Szenarien und die Aporien der Voraussicht." Archiv für Mediengeschichte 9 (2010): 3–21. Print.

Hornung, Alfred. "Postmoderne Bis Zur Gegenwart." *Amerikanische Literaturgeschichte*. Ed. Hubert Zapf. Stuttgart: Metzler: 2004. 306–386. Print.

Hulme, Mike. "Cosmopolitan Climates: Hybridity, Foresight and Meaning." *Theory, Culture & Society* 27.2-3 (2010): 267–276. Print.

———. "Four Meanings of Climate Change." *Future Ethics: Climate Change and Apocalyptic Imagination*. Ed. Stefan Skrimshire. London: Continuum Inter. Publis., 2010. 37–59. Print.

———. *Why We Disagree About Climate Change: Understanding Controversy, Inaction and Opportunity*. Cambridge; New York: Cambridge UP, 2009. Print.

Ingimundarson, Valur. "Die Kartierung Der Arktis: Bodenschätze, Großmachtpolitik und Multilaterale Governance." *Politik und Zeitgeschichte*, 5-6 (2011): 14–23. Print.

International Climate Justice Network. "Bali Principles of Climate Justice". EJNet. org. 29 August 2002. Web. 13 January 2014.

IPCC. *Climate Change 2007: Synthesis Report*. Valencia, Spain: N. p., 2007. Print.

———. *Climate Change 2013. The Physical Science Basis. Summary for Policymakers*. Fifth Assessment Report of the IPCC. Switzerland: N. p., 2013. Print.

———. *Emission Scenarios: Summary for Policymakers*. Geneva: N. p., 2000. Print.

———. "Organization." *IPCC-Intergovernmental Panel on Climate Change*. n.d. Web. 20 July 2012.

Jameson, Fredric. *Archaeologies of the Future: The Desire Called Utopia and Other Science Fictions*. London: Verso, 2005. Print

———. *Postmodernism, Or, The Cultural Logic of Late Capitalism*. Durham: Duke University Press, 1990. Print.

Johns-Putra, Adeline. "A New Critical Climate." *Symploke* 21.1-2 (2013): 1–5. Print.

———. "Ecocriticism, Genre, and Climate Change: Reading the Utopian Vision of Kim Stanley Robinson's Science in the Capital Trilogy." *English Studies* 91 (2010): 744–760. Print.

Johns-Putra, Adeline, and Adam Trexler. "Climate Change in Literature and Literary Criticism." *Wiley Interdisciplinary Reviews: Climate Change* 2 (2011): 185–200. Print.

Kerridge, Richard. "Ecothrillers: Environmental Cliffhangers." *The Green Studies Reader*. London; New York: Routledge, 2000. 242–249. Print.

———. "Narratives of Resignation: Environmentalism in Recent Fiction." *The Environmental Tradition in English Literature*. Ed. J. Parham. Aldershot: Ashgate, 2002. 87–99. Print.

Kilgore, De Witt Douglas. "Making Huckleberries: Reforming Science and Whiteness in Science in the Capital." *Configurations: A Journal of Literature, Science, and Technology*. 20.1/2 (2012): n.P. Print.

Killingsworth, Jimmie M., and Jaqueline Palmer. "Millenial Ecology. The Apocalyptic Narrative from Silent Spring to Global Warming." *Green Culture: Environmental Rhetoric in Contemporary America*. Ed. Carl Herndl and Stuart C. Brown. U of Wisconsin P, 1996. 21–45. Print.

Kingsolver, Barbara. *Flight Behavior. A Novel*. New York: Harper Collins, 2012. Print.

Kirksey, Eben, and Stefan Helmreich. "The Emergence of Multispecies Ethnography." *Cultural Anthropology* 25.4 (2010): 545–576. Print.

Kormann, Carolyn. "Scenes from a Melting Planet: On the Climate-Change Novel." *The New Yorker Blogs* 3 July 2013. Web. 10 Oct. 2013.

Kramb, Daniel. "Climate Change Fiction Melts Away Just When It's Needed." *The Guardian*. The Guardian. 18 Oct. 2012. Web. 22 Mar. 2014.

Kreibich, N., and H. Fechner. *Potentiale Ausgeschöpft und Hürden Überwunden? CDM und JI in der Ersten Kyoto-Verpflichtungsperiode*. Wuppertal Institut für Klima, Umwelt und Energie, 2013. Print.

Krugman, Paul. "Building a Green Economy." *The New York Times* 7 Apr. 2010. Web. 20 Feb 2014.

Krznaric, Roman. "Empathy and Climate Change: Proposals for a Revolution of Human Relationships." *Future Ethics: Climate Change and Apocalyptic Imagination*. Ed. Stefan Skrimshire. London: Continuum International Publishing, 2010. 153–173. Print.

Kuhn, Thomas. *The Structure of Scientific Revolutions*. Rev. 2nd ed. 1970. Chicago: U of Chicago P, 1962. Print.

Latour, Bruno. *Politics of Nature: How to Bring the Sciences into Democracy*. Trans. Catherine Porter. Harvard UP, 2004. Print.

———. *We Have Never Been Modern*. Harvard University Press, 1993. Print.

———. "Why Has Critique Run out of Steam? From Matters of Fact to Matters of Concern." *Critical Inquiry* 30 (2004): 225–248. Print.

Latour, Bruno, and Steve Woolgar. *Laboratory Life : The Construction of Scientific Facts*. 2nd ed. Princeton: Princeton UP, 1986. Print.

Lefebvre, Henri. *The Production of Space*. Trans. Donald Nicholson Smith. Oxford: Blackwell Publishing, 1991. Print.

Leggett, Jane A. "A U.S.-Centric Chronology of the International Climate Change Negotiations." *Congressional Research Service* (2010): n. p. Print.

Leggewie, Claus, and Harald Welzer. *Das Ende der Welt Wie Wir Sie Kannten. Klima, Zukunft und die Chancen der Demokratie*. Frankfurt a.M.: Fischer Taschenbuch Verlag, 2011. Print.

Leiserowitz, Anthony. "Before and After The Day After Tomorrow: A U.S. Study of Climate Risk Perception." *Environment* 46.9 (2004): 22–37. Print.

Lever-Tracy, Constance. *Confronting Climate Change*. New York: Routledge, 2011. Print.

Lewis, Michael. "In Nature's Casino." *NYTimes.com. The New York Times*, 26 Aug. 2007. Web. 25 Jan. 2014.

Lloyd, Saci. *The Carbon Diaries 2015*. New York: Holiday House, 2010. Print.

———. *The Carbon Diaries 2017*. New York: Holiday House, 2011. Print.

Lobo-Guerrero, Luis. *Insuring Security: Biopolitics, Security, and Risk*. New York: Routledge, 2011. Print.

Lovelock, James. *The Vanishing Face of Gaia: A Final Warning*. New York: Basic Books, 2010. Print.

Lübken, Uwe, and Christof Mauch. "Uncertain Environments : Natural Hazards, Risk and Insurance in Historical Perspective." *Environment and History* 17.1 (2011): 1–12. Print.

Luckhurst, Roger. "The Politics of the Network: The Science in the Capital Trilogy." *Kim Stanley Robinson Maps the Unimaginable*. Jefferson, N.C.: McFarland&Company, 2009. 170–181. Print.

Luke, Timothy W. *Ecocritique: Contesting the Politics of Nature, Economy, and Culture*. U of Minnesota P, 1997. Print.

Macfarlane, Robert. "The Burning Question." *The Guardian*. 24 Sept. 2005. Web. 26 Mar. 2012.

Markley, Robert. "'How to Go Forward': Catastrophe and Comedy in Kim Stanley Robinson's Science in the Capital Trilogy." *Configurations: A Journal of Literature, Science, and Technology* 20.1/2 (2012): N. pag. Print.

Mayer, Sylvia. "American Environmentalism and Encounters with the Abject: T. Coraghessan Boyle's A Friend of the Earth." *The Abject of Desire: The Aestheticization of the Unaesthetic in Contemporary Literature and Culture*. Ed. Konstanze Kutzbach and Monika Mueller. Amsterdam: Rodopi, 2007. 221–234. Print.

———. "Genre and Environmentalism: Octavia Butler's Parable of the Sower, Speculative Fiction, and the African American Slave Narrative." *Restoring the Connection to the Natural World*. Ed. Sylvia Mayer. Münster: Lit Verlag, 2003. Print.

———. "Literary Studies, Ecofeminism, and the Relevance of Environmentalist Knowledge Production in the Humanities." *Nature in Literary and Cultural Studies: Transatlantic Conversations on Ecocriticism*. Ed. Catrin Gersdorf and Sylvia Mayer Amsterdam: Editions Rodopi BV, 2006. 111–128. Print.

Mayer, Sylvia and Alexa Weik von Mossner, eds. *The Anticipation of Catastrophe: Environmental Risk in North American Literature and Culture*. Heidelberg: Universitätsverlag Winter, 2014 (American Studies Series). Print.

Mbembe, Achille. "Necropolitics." *Public Culture* 15.1 (2003): 11–40. Print.

McCarthy, Cormac. *The Road*. Vintage, 2007. Print.

McEwan, Ian. *Solar*. Random House UK, 2010. Print.

McKibben, Bill. "Introduction." *I'm With the Bears*. London;New York: Verso, 2011. 1–5. Print.

———. *The End of Nature*. Reprint. Random House Trade Paperbacks, 2006. Print.

McNeil, Jean. *Ice Lovers*. Toronto, Canada: McArthur & Co, 2009. Print.

Meadows, Donella and Dennis Meadows et al. *The Limits to Growth*. Signet, 1972. Print.

Mehnert, Antonia. "Climate Change Futures and the Imagination of the Global in *Maeva!* By Dirk C. Fleck." *Ecozon@* 3.2 (2012): 27–41. Print.

———. "Der Held und das Wetter: Literatur und Klimawandel." *Goethe Institut-Kultur und Klimawandel*. N. p., Nov. 2013. Web. 20 Feb. 2014.

Merker, Helmut. "Kapitalistisches Dreckwetter." *Der Tagesspiegel* 30 Jan. 1972. Web. 25 Jan. 2014.

Mills, William J. *Exploring Polar Frontiers: A Historical Encyclopedia*. ABC-Clio Inc, 2003. Print.

Mintz, S. "The Turner Thesis." *Digital History*. Web. 20 June 2012.

Mizruchi, Susan. "Risk Theory and the Contemporary American Novel." *American Literary History* 22.1 (2009): 109–135. Print.

Möllers, Nina. *Kreolische Identität. Eine Amerikanische "Rassengeschichte" Zwischen Schwarz und Weiß. Die Free People of Color in New Orleans*. Bielefeld: Transcript, 2008. Print.

Mollison, Bill. "About Permaculture." *Permaculture*. Web. 6 Nov. 2013.

Mortimer-Sandilands, Catriona. "Melancholy Natures, Queer Ecologies." *Queer Ecologies. Sex, Nature, Politics, Desire*. Ed. Catriona Mortimer-Sandilands and Bruce Erickson. Bloomington: Indiana UP, 2010. 331–358. Print.

Morton, Timothy. *Ecology Without Nature: Rethinking Environmental Aesthetics*. Cambridge, Mass.: Harvard UP, 2007. Print.

Moylan, Tom. *Scraps of the Untainted Sky: Science Fiction, Utopia, Dystopia*. Boulder: Westview Press, 2000. Print.

Moylan, Tom, and Raffaella Baccolini. "Introduction. Dystopia and Histories." *Dark Horizons: Science Fiction and the Utopian Imagination*. New York: Routledge, 2003. 1–13. Print.

Muir, John. *The Eight Wilderness-Discovery Books*. London: Diadem Books, 1992. Print.

Mulkern, Anne C. "Will Prize-Winning Novels Shift Attitudes on Global Warming?" *NYTimes.com*. The New York Times. 6 May 2011. Web. 21 Sept. 2012.

Murphy, Patrick. "Engineering Planets, Engineering Ourselves : The Ethics of Terraforming and Areoforming in an Age of Climate Change." *Journal of Ecocriticism* 1.1 (2009): 54–59. Print.

Murphy, Patrick D. "Community Resilience and the Cosmopolitan Role in the Environmental Challenge-Response Novels of Ghosh, Grace, and Sinha." *Comparative Literature Studies* 50.1 (2013): 148–168. Print.

Nixon, Rob. *Slow Violence and the Environmentalism of the Poor.* Cambridge, Mass.: Harvard UP, 2011. Print.

Nussbaum, Martha C. *Love's Knowledge: Essays on Philosophy and Literature.* Oxford University Press, 1992. Print.

———. *Poetic Justice: The Literary Imagination and Public Life.* Boston, Mass: Beacon Press, 2007. Print.

———. *Political Emotions: Why Love Matters for Justice.* Harvard UP, 2013. Print.

Obama, Barack. The White House, Office of the Press Secretary. "Inaugural Address." United States Capitol, 21 Jan. 2013. Web. 11 Feb. 2014.

Oliver-Smith, Anthony. "'What Is a Disaster?' Anthropological Perspectives on a Persistent Question." *The Angry Earth: Disaster in Anthropological Perspective.* Ed. Susannah M Hoffman and Anthony Oliver-Smith. New York: Routledge, 1999. 18–34. Print.

Organisation for Economic Co-operation and Development. *Cities and Climate Change.* Paris: OECD Publishing, 2010.

Oreskes, Naomi and Erik Conway. *Merchants of Doubt.* London: Bloomsbury, 2012. Print.

Pak, Chris. "Terraforming 101." *SFRA Review 302* (Fall 2012): 6–15. Print

Pelling, Mark. *Adaptation to Climate Change: From Resilience to Transformation.* London; New York: Routledge, 2011. Print.

Perrow, Charles. *Normal Accidents: Living with High-Risk Technologies.* Princeton,NJ: Princeton UP, 1999. Print.

Phillips, Jerry. "The Intuition of the Future: Utopia and Catastrophe in Octavia Butler's Parable of the Sower." *Novel* Spring/Summer (2002): 299–311. Print.

Plumwood, Val. "Nature in the Active Voice." *Australien Humanities Review* 46 (2009): n. pag. Print.

Potter, Emily. "Climate Change and the Problem of Representation." *Australien Humanities Review* 46 (2009): n. p. Web. 30 Apr. 2012.

Pratt, Mary Louise. *Imperial Eyes: Travel Writing and Transculturation.* London: Routledge, 2010. Print.

Price, Jenny. "Thirteen Ways of Seeing Nature in L.A." *The Believer.* April 2006. Web. 11 Oct. 2013.

Prettyman, Gib. "Living Thought: Genes, Genres and Utopia in the Science in the Capital Trilogy." *Kim Stanley Robinson Maps the Unimaginable.* Jefferson, N.C.: McFarland&Company, 2009. 181–204. Print.

Pulver, Simone, and Stacy VanDeever. "Thinking About Tomorrows: Scenarios, Global Environmental Politics, and Social Science Scholarship." *Global Environmental Politics* 9.2 (2009): 1–13. Print.

Raskin, Paul. *Great Transitions: The Promise and Lure of the Times Ahead.* Global Scenario Group, Stockholm Environment Institute, 2002. Print.
Ray, Sarah Jaquette. *The Ecological Other: Environmental Exclusion in American Culture.* Tucson: University of Arizona Press, 2013. Print.
Rich, Nathaniel. *Odds Against Tomorrow.* New York: Farrar, Straus & Giroux, 2013. Kindle file.
Robinson, Kim Stanley. *Forty Signs of Rain.* New York: Bantam, 2004. Print.
———. Fifty Degrees Below. New York: Bantam, 2005. Print.
———. *Sixty Days and Counting.* New York: Bantam Spectra, 2007. Print.
———. "Imagining Abrupt Climate Change. Terraforming Earth." *Amazon Shorts* (2005): 1–20. Print.
Rose, Deborah Bird et al. "Thinking Through the Environment, Unsettling the Humanities." *Environmental Humanities* 1 (2012): 1–5. Print.
Rose, Deborah Bird, and Thom Van Dooren. "Introduction." *Australien Humanities Review* 50 (2011): 1–4. Print.
Rosenlöcher, Thomas. "Hochwasser in Sachsen: Schon Wieder Flut, Christa." *Die Zeit.* 15 June 2013. Web. 17 June 2013.
Ross, Andrew. *Bird on Fire: Lessons from the World's Least Sustainable City.* Oxford University Press, 2011. Print.
———. *Science Wars.* Durham: Duke University Press, 1996. Print.
———. "The Social Claim on Urban Ecology." *The Nature of Cities.* Tucson: U of Arizona P, 1999. 15–33. Print.
Royal Society. "Geoengineering the Climate: Science, Governance, Uncertainty." *RS Policy Document.* London: 2009. Web. 5 February 2016.
Rushdie, Salman. *Imaginary Homelands: Essays and Criticism 1981-1991.* London: Vintage, 2010. Print.
Saenz-Arroyo, A., and et al. "Fishers' Anecdotes, Naturalists' Observations and Grey Reports to Reassess Marine Species at Risk: The Case of the Gulf Grouper in Gulf of California, Mexico." *Fish and Fisheries* 6 (2005): 121–133. Print.
Sarewitz, Daniel. "Science and Environmental Policy: An Excess of Objectivity." *Earth Matters: The Earth Sciences, Philosophy, and the Claims of Community.* Prentice Hall, 2000. 79–98 Print.
Sassen, Saskia. "Ethnicity and Space in the Global City: A New Frontier." *City and Immigration.* Barcelona: Centre de Cultura Contemporània de Barcelona, 1997. N. pag. Print.
Scherer, Glenn. "Christian-Right Views Are Swaying Politicians and Threatening the Environment." *grist* 28 Oct. 2004. Web. 20 June 2012.
Schlosberg, David. *Defining Environmental Justice: Theories, Movements and Nature.* Oxford: Oxford University Press, 2009. Print.
Schneider, Birgit. "Ein Darstellungsproblem Klimatischen Wandels? Zur Analyse und Kritik Wissenschaftlicher Expertenbilder und ihren Grenzen." *Kritische Berichte* 3 (2010): 80–90. Print.
Secretariat of the Convention on Biological Diversity. *Global Diversity Outlook 3.* Montreal, 2010. Print.

Selin, Henrik, and Stacy D. VanDeveer. "US Climate Change Politics and Policymaking." *Wiley Interdisciplinary Reviews: Climate Change* 2.1 (2011): 121–127.

Shapiro, Julie et al. *Climate Change and the Functioning of Water Rights : A Search of the Literature.* University of Colorado: Western Water Assessment Cooperative Institute for Research in Environmental, 2008. Print.

Shapiro, Michael J. "The Fear Dispositif." *Theory & Event* 14.1 (2011): n. pag. *Project MUSE.* Web. 12 Sept. 2012.

Shellnhuber, Hans Joachim. Interview by Katrin Elger und Christian Schwägerl. "Diktatur Des Jetzt." *Der Spiegel* 21 March 2011. Web. 20 March 2014.

Simpson, Helen. *In-Flight Entertainment.* New York: Vintage, 2012. Print.

Simpson, Lorenzo C. *Technology, Time, and the Conversations of Modernity.* New York: Routledge, 1994. Print.

Skrimshire, Stefan. "Introduction." *Future Ethics: Climate Change and Apocalyptic Imagination.* Ed. Stefan Skrimshire. London: Continuum Inter. Publis., 2010. 1–13. Print.

Slovic, Scott. *Going Away to Think: Engagement, Retreat, and Ecocritical Responsibility.* University of Nevada Press, 2008. Print.

Snow, C.P. *The Two Cultures. 1959.* Cambridge: Cambridge UP, 1993. Print.

Solnit, Rebecca. *A Paradise Built in Hell: The Extraordinary Communities That Arise in Disasters.* New York: Penguin Books, 2009. Print.

Soper, Kate. *What Is Nature: Culture, Politics and the Non-Human.* Malden, MA: Wiley-Blackwell, 1995. Print.

Soylent Green. Dir. Richard Fleischer. Perf. Charlton Heston, Edward Robinson. Metro-Goldwyn-Mayer, 1973. Film.

Spivak, Gayatri Chakravorty. "Extempore Response to Susan Abraham, Tat-Sion Benny Liew and Mayra Rivera." *Planetary Loves.* New York: Fordham UP, 2011. 136–149. Print.

———. "World Systems & the Creole." *Narrative* 14.1 (2006): 102–112. *CrossRef.* Web. 16 Feb. 2013.

Stafford, Michael. "The Heresy of Religious Opposition to Global Warming." *ABC Religion and Ethics* 19 Mar. 2012. Web. 20 June 2012.

Stein, Dana M. *Fire in the Wind.* Mustang, Oklahoma: Tate Publishing, 2010. Print.

Stillman, Peter. "Dystopian Critiques, Utopian Possibilities, and Human Purposes in Octavia Butler's Parables." *Utopian Studies* 14.1. (2003): 15–31. Print.

Sturgeon, Noël. *Environmentalism in Popular Culture : Gender, Race, Sexuality, and the Politics of the Natural.* Tucson: University of Arizona Press, 2009. Print.

Svendsen, Lars. *A Philosophy of Fear.* London: Reaktion Books, 2008. Print.

Swyngedouw, Erik. "Apocalypse Forever?: Post-Political Populism and the Spectre of Climate Change." *Theory, Culture & Society* 27.2-3 (2010): 213–232. Print.

Sze, Julie. "Boundaries of Violence: Water, Gender and Globalization at the US Borders." *International Feminist Journal of Politics* 9.4 (2007): 475–484. Print.

———. "From Environmental Justice Literature to the Literature of Environmental Justice." *The Environmental Justice Reader: Politics, Poetics, & Pedagogy.* Ed. Joni Adamson, Mei Mei Evans, and Rachel Stein University of Arizona Press, 2002. 163–180. Print.

———. "The Question of Environmental Justice." *Urban Climate Change Crossroads.* Ed. Richard Plunz and Maria Paola Sutto. Farnham, UK: Ashgate, 2008. 11–16. Print.

Szeman, Imre, and Maria Whiteman. "Future Politics: An Interview with Kim Stanley Robinson." *Science Fiction Studies* 31 (2004): 177–188. Print.

Szerszynski, Bronislaw, and John Urry. "Changing Climates: Introduction." *Theory, Culture & Society* 27.2-3 (2010): 1–8.

Theroux, Marcel. *Far North.* Faber and Faber, 2010. Print.

The Age of Stupid. Dir. Franny Armstrong. Perf. Pete Postlethwaite. Spanner Films, 2010. Film.

The Day After Tomorrow. Dir. Roland Emmerich. Perf. Dennis Quaid, Jake Gyllenhaal. Twentieth Century Fox, 2004. Film.

The National Science Foundation. "Antarctica. Past and Present" *United States in Antarctica.* nsf.gov. April 1997. Web. 20 March 2013.

The Yes Men. "All the News We Hope to Print." *The New York Times,* 12 Nov. 2008. Web. 2 June 2012. <http://www.spiegel.de/media/0,4906,19346,00.pdf>.

The Yes Men. "New York Post Special Edition." *Theyesmen.org.* The Yes Men Website, 20 Sept. 2009. Web. 20 Jan. 2014.

Thomashow, Mitchell. *Bringing the Biosphere Home: Learning to Perceive Global Environmental Change.* The MIT Press, 2003. Print.

Tomlinson, John. *Globalization and Culture.* Chicago: U of Chicago P, 1999. Print.

Trexler, Adam. "The Climate Change Novel: A Faulty Simulator of Environmental Politics." *Carnegie Council* 7 Nov. 2011. Web. 8 Mar. 2012.

———. *Anthropocene Fictions.* U of Virginia Press, 2015. Print.

Trojanow, Ilija. *EisTau.* München: Hanser, 2011. Print.

Tuan, Yi-Fu. *Space and Place.* Minneapolis: U of Minnesota Press, 1977. Print.

Turner, George. *Drowning Towers.* Avon Books, 1996. Print.

Uerz, Gereon. *ÜberMorgen: Zukunftsvorstellungen als Elemente der gesellschaftlichen Konstruktion der Wirklichkeit.* MÜnchen: Wilhelm Fink, 2006. Print.

United Nations. "Article 1: Definitions - Framework Convention on Climate Change." *Full Text of the Convention.* 21 March 1994. Web. 20 July 2012.

———. "Article 2: Objective - Framework Convention on Climate Change." *Full Text of the Convention.* 21 March 1994. Web. 20 July 2012.

US EPA. *Inventory of U.S. Greenhouse Gas Emissions and Sinks: 1990-2011.* Washington D.C.: US Environmental Protection Agency, 2013. Web. 14 Dec. 2013.

———. "U.S. and Global Temperature." *Climate Change Indicators in the United States.* August 2013. Web. 28 Feb. 2014.

———. "U.S. Greenhouse Gas Inventory Report." *Executive Summary.* National Greenhouse Gas Emissions Data. 12 April 2013. Web. 28 Feb. 2014.

Veeser, Aram. "Introduction." *The New Historicism.* Ed. Aram Veeser. New York; London: Routledge, 1989. ix–xvi. Print.

Wald, Priscilla. *Contagious: Cultures, Carriers, and the Outbreak Narrative.* Duke University Press, 2007. Print.

Walker, Brian, and David Salt. *Resilience Thinking: Sustaining Ecosystems and People in a Changing World.* Island Press, 2006. Print.

Wanzo, Rebecca. "Apocalyptic Empathy: A Parable of Postmodern Sentimentality." *Obsidian III* 6/7.2/1 (2005): 72–87. Print.

Wardekker, A.J.A., A.C. Petersen, and J.P. van der Sluijs. "Religious Positions on Climate Change and Climate Policy in the United States." Communicating Climate Change: Discourses, Mediations and Perceptions. University of Minho, Braga, Portugal, 19–20 November 2007. Address.

Weik von Mossner, Alexa. "Facing The Day After Tomorrow: Filmed Disaster, Emotional Engagement, and Climate Risk Perception." *American Environments: Climate-Cultures-Catastrophes.* Heidelberg: Universitätsverlag Winter, 2012. 97–115. Print.

———. "Troubling Spaces: Ecological Risk, Narrative Framing, and Emotional Engagement in The Age of Stupid." Ed. Mick Smith. *Emotion, Space, and Society* (2012): 108–116. Print.

Welzer, Harald. *Klimakriege: Wofür im 21. Jahrhundert Getötet Wird.* Frankfurt a. M.: Fischer, 2008. Print.

White, Jennifer Rose. "Trouble with Time. Contemporary American Literature and Environmental Crisis." Diss. Columbia University, 2009. Print.

Winterson, Jeanette. *The Stone Gods.* London: Penguin, 2008. Print.

Woolley, Agnes. "The Politics of Myth Making: 'Beasts of the Southern Wild.'" *openDemocracy.* 29. Oct. 2012. Web. 29 Nov. 2013.

Wynne, Bryan. "Strange Weather, Again: Climate Science as Political Art." *Theory, Culture & Society* 27.2-3 (2010): 289–305. Print.

Yaeger, Patricia. "Beasts of the Southern Wild and Dirty Ecology." *Southern Spaces* 13 Feb. 2013. Web. 29. Nov. 2013.

Yates, Joshua. "An Interview with Ulrich Beck on Fear and Risk Society." *The Hedgehog Review* Fall 03 (2004): 96–107. Print.

Young, Iris. *Justice and the Politics of Difference.* Princeton UP, 1990. Print.

Yusoff, Kathryn. "Biopolitical Economies and the Political Aesthetics of Climate Change." *Theory, Culture & Society* 27.2-3 (2010): 73–99. Print.

Zapf, Hubert. "Literary Ecology and the Ethics of Texts." *New Literary History* 38.4 (2008): 1–16. Print.

———. "New Directions in American Literary Studies. Ecocriticism and the Function of Literature as Cultural Ecology." *English Studies Today. Recent Developments and New Directions.* Ed. Ansgar Nünning and Jürgen Schlaeger. Trier: WVT, 2008. 139–164. Print.

———. "The State of Ecocriticism and the Function of Literature as Cultural Ecology." *Nature in Literary and Cultural Studies: Transatlantic Conversations on Ecocriticism.* Ed. Catrin Gersdorf and Sylvia Mayer. Amsterdam: Editions Rodopi BV, 2006. 49–71. Print.

Ziser, Michael. "Home Again: Peak Oil, Climate Change, and the Aesthetics of Transition." *Environmental Criticism for the Twenty-First Century.* Ed. Stephanie LeMenager, Teresa Shewry, and Ken Hiltner. New York: Routledge, 2011. 181–196. Print.

Ziser, Michael, and Julie Sze. "Climate Change, Environmental Aesthetics, and Global Environmental Justice Cultural Studies." *Discourse* 29.2-3 (2009): 384–410. Print.

Žižek, Slavoj. "The Situation is Catastrophic." *Lacan.com.* N.d. Web. 5 Jan. 2014.

Index

A. Mehnert, *Climate Change Fictions*,
DOI 10.1007/978-3-319-40337-3

The manufacturer's authorised representative in the EU is Springer Nature Customer Service Centre GmbH, Europaplatz 3, 69115 Heidelberg, Germany. If you have any concerns regarding our products, please contact ProductSafety@springernature.com

Printed and bound by CPI Group (UK) Ltd, Croydon, CR0 4YY
30/09/2026
02225565-0009